Ultrastructure of the Conjunctival Epithelium

Developments in Ophthalmology

Vol. 19

Series Editor

W. Straub, Marburg a. d. Lahn

KARGER

Basel · München · Paris · London · New York · New Delhi · Singapore · Tokyo · Sydney

Ultrastructure of the Conjunctival Epithelium

Klaus-Peter Steuhl, Tübingen

53 figures and 6 tables, 1989

KARGER

Basel · München · Paris · London · New York · New Delhi · Singapore · Tokyo · Sydney

Developments in Ophthalmology

Continuation of 'Bibliotheca Ophthalmologica', 'Advances in Ophthalmology', and 'Modern Problems in Ophthalmology'.
Founded 1926 as 'Abhandlungen aus der Augenheilkunde und ihren Grenzgebieten' by *C. Behr,* Hamburg and *J. Meller,* Wien.
Former Editors: *A. Brückner,* Basel (1938–1959); *H. J. M. Weve,* Utrecht (1938–1962); *H. M. Dekking,* Groningen (1954–1966); *E. B. Streiff,* Lausanne (1954–1979); *J. François,* Gand (1959–1979); *J. van Doesschate,* Utrecht (1967–1971); *M. J. Roper-Hall,* Birmingham (1966–1980); *H. Sautter,* Hamburg (1966–1980).

Bibliographic Indices
This publication is listed in bibliographic services, including Current Contents® and Index Medicus.

Drug Dosage
The author and the publisher have exerted every effort to ensure that drug selection and dosage set forth in this text are in accord with current recommendations and practice at the time of publication. However, in view of ongoing research, changes in government regulations, and the constant flow of information relating to drug therapy and drug reactions, the reader is urged to check the package insert for each drug for any change in indications and dosage and for added warnings and precautions. This is particularly important when the recommended agent is a new and/or infrequently employed drug.

Printed by Konkordia Druck GmbH, Bühl/Baden (FRG)
ISBN 3–8055–4943–1

Contents

Preface VII
Acknowledgements IX
Abbreviations X

1. Introduction 1
1.1. The Tear Film 1
1.1.1. The Mucin Layer 2
1.1.2. The Aqueous Layer 2
1.1.3. The Lipid Layer 6
1.2. The Conjunctival Epithelium 6
1.2.1. Surface Structure of Conjunctiva 8
1.2.2. Light and Transmission Electron Microscopy of Conjunctiva 9
1.2.3. Functions of the Conjunctival Epithelium 10

2. The Question 12

3. Materials and Methods 14
3.1. Materials 14
3.1.1. Chemicals 14
3.1.2. Solutions 15
3.2. Methods 16
3.2.1. Transmission Electron Microscopy 16
3.2.2. Scanning Electron Microscopy 17
3.2.3. Carbonic Anhydrase Reaction 17
3.2.4. Peroxidase Reaction 17
3.3. Animal Conjunctiva 18
3.3.1. Albino Rabbits 18
3.3.2. Cynomolgus Monkeys 20
3.4. Human Conjunctiva 22
3.4.1. Healthy Conjunctival Tissue 22
3.4.2. Pathologically Altered Conjunctival Tissue 22

4. Results 28
4.1. Animal Conjunctiva 28
4.1.1. Ultrastructure of Conjunctival Epithelium in Albino Rabbits and Cynomolgus Monkeys 28

4.1.2. Resorption Processes in Rabbit and Monkey Conjunctival Epithelium 39
4.1.3. Effect of an Antiglaucomatous Agent (Pilocarpine) on the Conjunctival Ultrastructure of Cynomolgus Monkeys 42
4.1.4. Corneal Reepithelialization at the Rabbit Limbus: Ultrastructure and Histochemistry 44
4.2. Human Conjunctiva 50
4.2.1. Ultrastructure of Healthy Conjunctival Epithelium: Age-Related and Regional Differences 50
4.2.2. Ultrastructure of Pathological Conjunctival Tissue 65

5. Discussion 72
5.1. Morphology and Function of Rabbit and Monkey Conjunctival Epithelium 72
5.1.1. Secretion of Mucoid Substances 72
5.1.2. Defense, Regulation, Self-Ablution 74
5.1.3. Resorption 74
5.2. Age-Related Morphological Changes in Human Conjunctiva 78
5.3. Effect of Pathological Processes on Cell Distribution in Human Conjunctiva . . . 82
5.3.1. Conjunctival Biopsy or Impression Cytology? 82
5.3.2. Dry Eye (e.g. Sjögren's Syndrome) 82
5.3.3. Ultrastructure of the Conjunctival Epithelium after Lime Burn 84
5.3.4. Effect of Antiglaucomatous Drugs on the Ultrastructure of the Conjunctival Epithelium 85

6. Conclusion 87

7. References 89

Subject Index 100

Preface

The conjunctival epithelium has often been neglected in the past because of its supposed insignificance. Only recently has it become an acceptable subject of ophthalmological congresses. The conjunctiva itself was regarded as a relatively undifferentiated entity; only its goblet cells attracted much scientific interest because of their contribution to the mucin component of the tear film.

This disinterest has evaporated rapidly, however, with current advances in understanding of the physiology and pathology of the precorneal tear film. New light has been cast on the role of conjunctival secretory mechanisms. Deeper insight into immunological cause-and-effect relationships has led to fresh explanations for the etiology and pathogenesis of conjunctival diseases. Moreover, other components of the tear film have been identified (e. g. proteolytic and glycolytic enzymes, various antibodies, glycoproteins, glucose, calcium) which certainly participate in the secretory activity of conjunctival cells and have their morphological counterparts as well.

Given the numerous external factors to which the conjunctiva and the outer eye are exposed and the variety of components found in the precorneal tear film, the author of this study proposes that several functional systems, whose structures are not yet fully understood, are involved in conjunctival secretion.

A significant merit of this book is that it differentiates histologically and electron microscopically among conjunctival epithelial cell types and then traces different functions to them on the basis of morphological traits. The interesting insight that different functional systems have their respective morphological basis will allow the clinician a better classification of regonizable changes of the outer eye.

Finally, the author demonstrates that provocation of the outer eye or the conjunctiva by chemical burns or drugs causes a transformation in cell ultrastructure, resulting in functional impairment.

This book offers not only important insights into the cellular structure of normal and diseased conjunctiva but also new information for the interpretation and understanding of various diseases. Its elaboration of different conjunctival systems provides the clinician with a better interpretation of individual symptoms. Therefore, this book deserves a wide distribution.

Tübingen, Spring 1989 *Hans-Jürgen Thiel*

Acknowledgements

My thanks go to Prof. H.-J. Thiel, MD (University Eye Clinic, Tübingen), without whose advice and vigorous support this study could not have been completed. I acquired the theoretical foundations for this work under the direction of Prof. E. Lütjen-Drecoll, MD, and Prof. Dr. Dr. J. W. Rohen, MD (Anatomical Institute of the University of Erlangen-Nürnberg). Their support and valuable suggestions continued even after my anatomical training. Prof. K. Dietz, MD (Institute of Medical Biometry, University of Tübingen), and M. Fahle, MD (University Eye Clinic, Tübingen), made many helpful suggestions concerning statistical evaluation of the data. Mrs. Gutbrod, Mrs. Kittel and Mrs. Liebig, medical technicians at the University Eye Clinic in Tübingen, assisted me in the experimental work. My thanks to Mr. J. Orlewski and to Mr. M. Nicolescu for the graphic presentation of the numerical data. The photographs and text were prepared in exemplary fashion by the photographic department of the University Eye Clinic and Mr. Thomas Rice. My thanks, too, to all my colleagues at the University Eye Clinic in Tübingen who contributed to the successful completion of this work. My wife Danielle was a pillar of support with her unfailing understanding for my work.

I am indebted to Pharm-Allergan (Karlsruhe, FRG) for financial assistance regarding publishing costs.

Abbreviations

Cat. Ex.	Cataract extraction
HP	Horseradish peroxidase
KP	Keratoplasty
MPS-Granules	Mucopolysaccharide granules
rER	Rough endoplasmatic reticulum
SEM	Scanning electron microscope
TEM	Transmission electron microscope

1. Introduction

A functional lid apparatus is indispensable for corneal transparency [158]. The numerous glands of the lid complex produce the precorneal tear film, a highly specialized, composite layer of fluid. Manifold qualitative and quantitative disturbances of this protective layer are possible, with results ranging from subjective impairment to blindness. Damage to the precorneal tear film results not only from positional anomalies, scarring, lid paralysis, infectious or age-related lacrimal gland atrophy, contact lenses, and environmental conditions, but frequently from conjunctival diseases as well. Thus the conjunctiva not only has mechanical and sensory functions [136, 153, 163] and the capability of resorption [105], but also makes an important contribution to the integrity of the tear film.

1.1. The Tear Film

The tear film can be divided into two parts, one preocular, the other precorneal. The preocular tear film includes the tears at the lid margins, on tarsal and bulbar conjunctiva, and at the superior and inferior conjunctival folds. The part lying directly on the cornea is known as the precorneal tear film.

Tear film volume is normally 7–9 μl [168]. About 1–2 μl of tear fluid are excreted per minute [122, 130, 143, 174]. Peripheral, sensory stimulation (especially of the cornea), retinal stimulation and psychic stimuli can increase this amount 20–30 times [87]. Such 'reflex' secretion is different from the 'basic' secretion described above in that it is governed by nerves, and its volume thus depends largely on the lacrimal gland's secretion. Basic secretion, on the other hand, is produced by glands of the lid complex, which regularly secrete fluid in constant amounts [86, 87]. The tear film is thickest (9 μm) directly after a blink and thinnest (ca. 4 μm) some 30 s later [132].

1.1.1. The Mucin Layer

Wolf [204] identified three layers in the precorneal tear film (fig. 1). The mucin layer, which adheres directly to the membrane of the epithelial surface cells, is about 0.8 μm thick on the cornea and ca. 1.4 μm thick on the conjunctiva [127]. This mucin layer consists of glycoproteins [38, 62, 78, 125]. The mucins are produced not only by goblet cells, but also by elements of the lacrimal gland [8, 83] and by a second mucus-secreting system of conjunctiva [35, 65, 159]. The mucin layer's function is to keep the corneal surface smooth and mediating water-bonding with its hydrophilic side, thus enabling moisturization of the hydrophobic corneal surface by the aqueous layer of the tear film [74]. Above this basally adsorbed, thin mucin layer is a more concentrated layer of coacervated glycoprotein complexes which merge into an even further diluted mucin solution towards the tear film surface (fig. 1).

1.1.2. The Aqueous Layer

The middle layer of the tear film is called the 'aqueous layer'. It is some 8 μm thick, constitutes somewhat more than 98% of the precorneal tear film, and consists of a watery solution of electrolytes and organic substances [74, 199]. Also present in the aqueous layer are living and desquamated dead cells [131].

The tear film's pH ranges from 7.14 to 7.82 [26], with reflex secretion leading to a primarily basic value. The tear fluid's osmotic pressure is about 305 mosm/kg [190], corresponding to a sodium chloride equivalent of 0.93–1.01%. Continuous lid closure reduces the osmolarity of the tear film to 282–288 mosm/kg, since the aqueous layer cannot evaporate. The partial oxygen pressure of the tear film ranges from 140–160 mm Hg. At least 74 mm Hg PO_2 are required for adequate corneal metabolism [73].

Numerous proteins have been isolated from the tear fluid, most of which were found dissolved in the tear film's aqueous layer. Protein concentration is normally 20 g/l, but drops to 3–7 g/l following stimulation of tear flow (so-called 'reflex secretion') [199]. The proteins of human tear fluid contain a specific tear albumin (prealbumin) as well as immunoglobulins, infection mediators (e.g. complement), plasminogen activators, proteases, and antibacterially effective proteins such as lysozyme, β-lysin and lactoferrin [82]. Also detectable are histamine as a biogenic amine and prostaglandines as derivatives of long-chain unsaturated fatty acids [3, 33]. Taken as a whole, these active substances have numerous functions; above

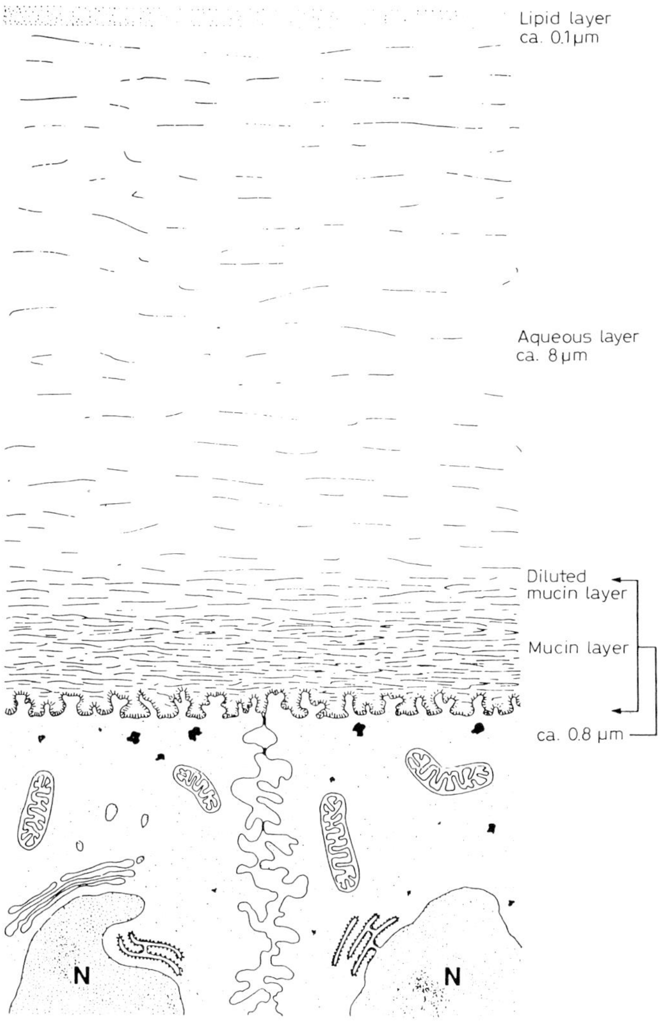

Fig. 1. Schematic view of the precorneal tear film.

all, by reducing surface tension they improve moisturization of the corneal surface and make a uniform distribution of the tear film possible. Other important functions involve the regulation of osmotic pressure and pH as well as the management of infections [7, 27].

1.1.2.1. Antimicrobial Proteins in the Aqueous Layer

The tear fluid contains all components necessary for the control of infectious processes [173]. Lysozyme is an unspecific bacteriolytic protein concentrated much more highly in the tear fluid (1–2 g/l) than elsewhere in the body [15, 171]. Reflex lacrimation increases the lysozyme content of the tears by as much as 100% [184]. Thus lysozyme is secreted for the most part by the lacrimal gland and Krause's accessory tear glands [56]. Many authors in fact regard the lysozyme content of the tear fluid as a test of the lacrimal gland's proper functioning [81]. Moreover, β-lysin, a protein which in contrast to lysozyme is heat-labile, contributes to the bacteriolytic activity of the tear fluid [44]. Another protein with bacteriostatic effect is lactoferrin. With its ability to bind Fe^3 300 times more firmly than transferrin, it reduces the content of metal ions essential to bacterial metabolism in the tear fluid [21, 55]. In addition, lactoferrin also seems to have a direct bacteriostatic effect on specific strains of bacteria [11]. Interferon can also be induced in the lacrimal fluid as an unspecific, antivirally effective protein [28].

1.1.2.2. Antibodies in the Aqueous Layer

All immunoglobulins are present in the tear film; their concentration is higher during basic secretion than after reflex tear secretion (table 1) [19, 117]. This shows that they make their way into the tear film primarily by means of the conjunctiva and only to a lesser extent through the lacrimal gland. Experiments show that specific antibodies are quick to appear in the lacrimal fluid of immunized animals [164, 180].

Predominant among the immunoglobulins of the lacrimal fluid is immunoglobulin A (IgA). As with other secretions of mucous membranes,

Table 1. Immunoglobulins in the tear fluid and serum [117]

Substance	Tear film mg/100 ml	Serum mg/100 ml	Proportion tear film/ serum
Total protein	800	6,500	1:8
IgG	14	1,000	1:70
IgA	17	170	1:10
IgM	< 5	100	<1:18
IgD	< 1	11	1:11
IgE	250 ng/ml	2,000 g/ml	1:8

two IgA molecules (MW 300,000) are bound to one another in the lacrimal fluid by a secretory piece (MW 85,000) [49, 118]. This secretory piece is synthesized in both the lacrimal gland and conjunctival epithelium, but not in the corneal epithelium [47, 106]. IgA itself is produced by the plasma cells of the conjunctival lymphatic tissue. Its secretory piece protects it on the epithelial surface from enzymatic destruction [48]. The immunoglobulins protect the mucous membrane surface through agglutination of bacteria; they also inhibit the adhesion of bacteria to epithelial cells, neutralize viruses, and make antibody-dependent cytotoxic reactions possible [123, 201].

1.1.2.3. Other Biologically Active Components of the Aqueous Layer

Some 30% of tear protein is constituted by a tear-specific prealbumin not found in blood serum [16, 88]. In all likelihood, this tear-specific prealbumin is important for the stability and spread of the precorneal tear film. Proteases are also present within the aqueous layer of the tear film [195]; their concentration can be of diagnostic value in various pathological conditions [25, 84], and they exercise an important function in regulating biological processes [126]. This is also true of antiproteases, although they are much less concentrated in tear fluid than in blood plasma [151, 209]. Above all, two processes in the tear film area are regulated by the interplay of proteases and antiproteases, namely infection and blood coagulation with fibrinolysis [13, 160, 210].

Two further substances, namely histamine and prostaglandine, play a major role in the control of external eye infections. Histamine, a mediator of allergic reactions, normally has a concentration of ca. 10 mg/ml in the tear fluid [3]. This concentration increases markedly in allergic reactions of the external eye and can be used in this context as a diagnostic indicator [91]. The same holds true for prostaglandins, which have likewise been found in the tear fluid [33, 57] and lead to conjunctival injection, chemosis and intensified mucus production upon superficial application [4, 39].

Complement activity in the human tear fluid is a subject of controversy. Whereas Yamamoto and Allansmith [205] established in 10 test persons that both classical and alternative complement activity are possible in the tear fluid, Kijlstra et al. [95] maintain that factors predominant in the tear fluid work to suppress complement-dependent cell lysis. Moreover, enzymes are found in the tear fluid which actively participate in energy-producing metabolic processes [198]. Also present are metabolites such as lactate, urea and catecholamines (dopamine, adrenaline and noradrenaline) [199].

1.1.2.4. Cellular Components of the Aqueous Layer

The preocular tear film also contains cells; these originate on the one hand in the corneal and conjunctival epithelium and the subconjunctival capillaries and on the other in lymphatic tissue associated with the conjunctiva. 80% of leukocytes and desquamated epithelial cells are, however, degenerated or already dead [131, 133]. Lymphocytes, plasma cells and leukocytes appear in greater numbers in the tear fluid following injury to or infection of the external eye; polymorphonuclear leukocytes predominate in these acute processes [185].

1.1.3. The Lipid Layer

The aqueous layer of the tear film is covered by an approx. 0.1-μm thick lipid layer which constitutes about 1% of the tear film's total thickness [89, 134]. Without this lipid layer, evaporation of the tear film's aqueous layer would increase 15-fold [77, 121]. The lipids of the tear film originate in Meibomian, Zeiss' and Moll's lid glands [87]. They consist of wax (35%), cholesterol esters (32%), phospholipids (16%), neutral fats, i.e. triglycerides (4%), and other lipids (13%) [10, 128]. Their melting point is 35 °C, and they are completely liquefied at 40 °C [23]. In the liquid state, the polar elements of this lipid film – i.e. the phospholipids and cholesterol esters – are the first to spread out over the aqueous layer of the tear film after each lid blink. Thereafter the less polar lipid components form a molecular layer, resulting in a so-called 'duplex layer' [194]. The stability of this lipid layer is increased by interaction between the lipid film's polar elements and the aqueous layer's mucin and protein molecules [76]. By causing water droplets to run off, this hydrophobic tear film coating gives additional protection against droplet infection.

1.2. The Conjunctival Epithelium

The multilayered, keratinizing lid epithelium merges into the multilayered, nonkeratinized conjunctival epithelium at the lid margins near the openings of the Meibomian glands. The hydrophobic, fatty secretion of these glands also limits the spread of the tear film (fig. 2).

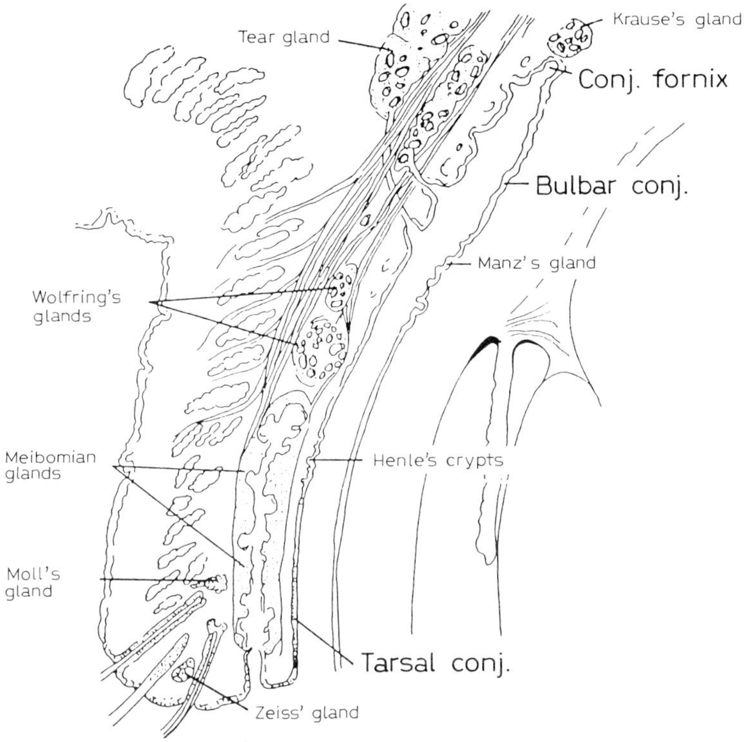

Fig. 2. Cross-section of upper lid conjunctiva, showing the glands responsible for tear secretion.

The lid margins are covered anteriorly by a dry, keratinized epithelium and posteriorly by a moistened, nonkeratinized epithelium. Near the posterior lid margin, the epithelium folds back almost at a right angle to cover the tarsus of both lids. The conjunctiva then forms the inferior and superior fornices, constituted by the transition of the tarsal conjunctiva onto the outer surface of the globe. The conjunctival epithelium ends on the anterior surface of the globe near the limbus, where the flat, multilayered, nonkeratinized squamous epithelium of the bulbar conjunctiva merges seamlessly into the corneal epithelium. This construction results in a mucous-membrane-lined sac covering the globe with its posterior portion but anteriorly open to the environment where the lids part. In this way the conjunctiva unites the eyelids with the globe. Due to the sac construction, the eyelids and the globe can move independently of each other.

On the average, the superior fornix is 13 mm from the lid margin, the inferior fornix 9 mm and the lateral fornix 5 mm [157]. The conjunctival sac extends medially not quite to the canthus; it is bounded there by the rudimentary human nictitating membrane, the so-called 'semilunar fold'. The conjunctiva is usually divided into (1) a marginal section from the Meibomian gland exits to 2 mm below the lid edge, (2) a tarsal section covering the superior and inferior tarsus, (3) a fornical section coating the conjunctival fold, and (4) a bulbar section contiguous with the globe. The limbal epithelium, which constitutes a morphological and histochemical unit, is usually discussed separately [97, 181].

The marginal section of the conjunctiva cleanses and moistens the bulbar and corneal epithelia during blinking. In addition, it distributes the secretion of the Meibomian glands.

A highly specialized system of glands evolves from the conjunctival epithelium; in essence, their secretion preserves the transparency of the cornea (fig. 2) [158]. As already mentioned, lipid secretion is a function of the sebaceous Meibomian glands (ca. 25 in the upper lid tarsus, ca. 20 in the lower lid tarsus), the Zeiss' glands (on the lid margin) and the Moll's glands at the roots of the eyelashes. Secretion of the tear film's aqueous layer, on the other hand, is a joint function of the lacrimal gland, Krause's subconjunctival, accessory lacrimal glands (in the superior and inferior conjunctival fold) and Wolfring's glands (at the upper and lower margins of the superior and inferior tarsus). The components of the mucin layer are produced chiefly by the conjunctiva's goblet cells. Conglomerates of these cells form 0.5-mm large, saccular invaginations – so-called Henle's crypts – in the region of the tarsal conjunctiva. Other intraepithelial collections of goblet cells, known as Manz's glands, are located approx. 3–7 mm nasally from the corneal limbus on the bulbar conjunctiva [94]. This glandular system of conjunctiva and eyelids – with the exception of the lacrimal gland – produces the tear film's basic secretion. It is not nervally regulated. All three parts of the tear film are formed during this basic secretion [87] (fig. 2).

1.2.1. Surface Structure of the Conjunctiva

Due to its numerous intraepithelial glands, the surface profile of the tarsal conjunctiva is particularly rugged, with many gullies and openings. Small foreign bodies can become lodged in these 'crypts' without leading to irritation of the corneal epithelium: they become coated with mucus in the crypts and are then transported toward the draining tear passages by the medially directed wiping motion of each lid blink [156]. The surface structures just described are accessible to biomicroscopic examination in living patients.

Other characteristic surface details of the conjunctiva appear in the scanning electron microscope (SEM). The conjunctival surface cells are for the most part hexagonal and are completely covered with microvilli. Some of these microvilli coalesce into so-called microplicae, while others divide [64, 196]. The diameter of the microvilli is 0.5 μm, their height between 0.5 and 1 μm. These microscopic protuberances in the epithelial surface cells of the conjunctiva stabilize and anchor the tear film on the one hand while enlarging the resorbent surface of the epithelium on the other [36, 105, 178]. The SEM also makes it possible to distinguish among light, middle-colored and dark cells. The lighter-colored cells are the most frequent and have fewer but longer microvilli, while the middle-colored and dark cells have relatively many but somewhat more compact microvilli [63]. Distributed over the entire epithelial surface are intercellular openings 1–3 μm in size, primarily at points where three or four epithelial cells meet. These are the openings of intraepithelially located goblet cells [63, 139]. Depending on the secretory cycle of each goblet cell, the openings are either empty in the SEM or contain plugs of mucus. The goblet cells are not recognizable with the SEM until the secretory cycle has begun [66]. Larger openings (10–60 μm in diameter) correspond to the entryways of the epithelial crypts mentioned above. Their borders and inner surfaces are usually coated with microvilli [63, 160]. The distribution of these crypts corresponds to Kessing's experimental findings [94].

1.2.2. Light and Transmission Electron Microscopy of the Conjunctiva

The conjunctival epithelium is multilayered. Its two to five cell layers are covered at the margins and in the limbal area by a flat surface cell layer. The pronounced flattening of the epithelial surface cells in these areas is regarded as an adaptation to increased mechanical pressure on these epithelial zones [142]. In bulbar and tarsal conjunctiva, the epithelial surface cells are more or less cubical [30, 188]. Both epithelia undergo a transition into the multilayered cylindrical epithelium of the conjunctival fornix [172, 202].

The epithelium grows markedly flatter with increasing age, and hyaline substances are deposited intracellularly. There are also fewer microvilli [1].

Highly expandable spaces are present between the epithelial cells of the conjunctiva. These intercellular spaces contain small cell extensions and are firmly closed surfacewards by tight junctions. Basally at the border between epithelium and stroma, however, there are no such junctions, so that the intercellular spaces here are often wide open. The high elasticity and pliability of these intercellular spaces of the conjunctival epithelium increases their ability to sustain mechanical stress on the one hand and to serve in resorption processes on the other [161, 178].

Most previous ultrastructural studies of the conjunctiva have analyzed animal rather than human material. It has been customary in such studies to contrast goblet cells with the more or less undifferentiated epithelial cells [172, 186, 200, 202]. Although Latkovic and Nilsson [104 a–d] have made numerous studies of guinea pig conjunctival epithelium with the transmission electron microscope (TEM), giving attention to specific epithelial regions, their reports make no further differentiation of epithelial conjunctival cells. They refer only to the presence of different cell organelles.

Somewhat more detailed morphological conceptions exist concerning the surface epithelium of the conjunctival limbus. This is an approx. 1.5-mm wide epithelial zone covering the area where cornea and sclera merge (fig. 4). Here radially arranged concentrations of fibrovascular tissue, the 'palisades of Vogt', are present. The limbal epithelium is up to 15 cells thick in the depressions between the palisades and a maximum of 3 cell layers thick on the palisades themselves [59]. The general opinion is that this 'reservoir' of epithelial cells serves to cover corneal epithelial defects quickly through migration [71, 97]. Epithelial surface cells in the limbus region of rabbits and primates are flat near the cornea, but gradually assume a cubical shape as they approach the bulbar conjunctiva. The epithelium in this region contains no goblet cells. Numerous melanocytes and Langerhans' cells are present near the basal epithelial cell layer. The cytoplasm of the epithelia contains many mitochondria and, to a lesser extent, rough endoplasmatic reticulum (rER) with scattered Golgi complexes. Tonofilaments and desmosomes are also present in abundance [12, 155, 181]. Histochemically, this epithelial region is clearly distinguishable from the neighboring epithelia [97, 111].

1.2.3 Functions of the Conjunctival Epithelium

In addition to the mechanical and sensory functions already mentioned, two other important functions of the conjunctiva have emerged from various experimental data.

1.2.3.1. Secretion

To date the goblet cells and the so-called second mucus-secreting system of conjunctiva have been identified as sources of important conjunctival secreta. The distribution of the goblet cells and the chemical composition of their secretion have been repeatedly studied [66, 94, 119, 124]. It has been shown that goblet cells have a characteristic pattern of distribution within the human conjunctival epithelium. They are most frequent in the nasal conjunctival area and only rarely present where the conjunctiva is in direct contact with the environment. Moreover, they are not found in conjunctival epithelium where it is elevated due to subconjunctival lymph follicles. This indicates an increased sensitivity of this cell type to mechanical stress [94]. An increased frequency of goblet cells in childhood was postulated by Kessing [94], but this has been challenged by Marquardt et al. [114], who emphasized the considerable interindividual variation in goblet cell density and have shown that goblet cells thin out slightly, if at all, only after the 60th year of life. Wanko et al. [200] and Takakusaki [188] were the first to show that not only goblet cells but also other epithelial cells have membrane-bound vesicles. Srinivasan et al. [176] demonstrated in histochemical studies of the conjunctiva that these vesicles are not lysosomes but contain glycoproteins. These findings were taken up by other authors, and it proved possible to identify a so-called second mucus-secreting system of the conjunctiva [35, 67, 159, 179].

1.2.3.2. Resorption

Studies with radioisotopes have shown that in patients with occluded efferent tear ducts ca. 30% of an amount of technetium placed in the conjunctival sac is resorbed systemically within 15 min [175]. The importance of the conjunctiva for the adsorption of superficially applied medications has been recently underscored experimentally by Ahmed and Patton [6]. It has also been shown by means of incorporation of latex particles and bacteria into epithelial cells that the cells of the conjunctival epithelium are capable of true phagocytosis and that latex particles need only 24 h to migrate through the epithelial surface layer [105, 208].

2. The Question

Until recently, the conjunctiva was invariably regarded as a relatively uniform tissue. Only the goblet cells aroused a certain amount of scientific interest due to their importance for the mucin layer of the tear film. Newer studies have shown, however, that the conjunctival epithelium is much more involved in the composition of the tear film than was previously supposed. It is probable that not only the mucins of the conjunctiva's above-mentioned second secretory system but also other tear film components are not released into the tear film by the lacrimal gland alone. Many substances – e.g. proteolytic and glycolytic enzymes, various antibodies, glycoproteins, glucose, and calcium – are less concentrated in the lacrimal fluid produced after reflex stimulation than in the tears of the basic secretion [19, 38, 165, 199]. As already mentioned reflex secretion affects only the volume secreted by the lacrimal gland, so that the above substances are probably of conjunctival origin.

Moreover, the reduced osmolarity of the reflex tears indicates that participation of the conjunctiva is essential to the secretion of molecular components of the lacrimal fluid [190]. Histochemical and immunological studies have shown that the conjunctival epithelium is capable of synthesizing various proteins whose activity is partly enzymatic [103, 106, 160]. With the aid of tissue cultures, it has also been shown that cells of the conjunctival epithelium can produce and release cytokines [61, 170]. All these findings indicate that the surface epithelial cells of the conjunctiva not only perform mechanical functions but also make an important contribution to the physiology of the tear film. The morphological significance of these multifarious functions has not yet been explained.

The task of this study will therefore be first of all to describe the ultrastructure of the conjunctiva in detail and to identify its regional characteristics. This will be done first with an animal model and then with human research material. In addition, certain functional aspects of the conjunctiva in animal experiments will be presented and their ultrastructural foundations will be described.

Because numerous conjunctival diseases, 'dry eye', and locally applied medications have been suspected of impairing both the conjunctival epithelium and the tear film, a study of morphological changes after lime burns, Sjögren's syndrome (dry eye) and the local application of β-blockers and pilocarpine will follow. The results will be compared to those of conjunctival specimens from healthy subjects.

Finally, an attempt will be made in light of this morphological study of the conjunctiva (including age-related changes) to establish a basis for better recognizing and classifying pathological conditions.

3. Materials and Methods

3.1. Materials

3.1.1. Chemicals

Acetone	p.a.	Merck, Darmstadt
Ammonium sulfide	p.a.	Merck, Darmstadt
Ethanol	p.a.	Merck, Darmstadt
Lead citrate	p.a.	Merck, Darmstadt
Cacodylic acid sodium salt		Serva, Heidelberg
Diaminobenzidine		Serva, Heidelberg
Diamox®		Cyanamid-Novalis, Wolfratshausen
Disodium hydrogen phosphate	p.a.	Merck, Darmstadt
Epon		Serva, Heidelberg
Epon accelerator DMP-30		Serva, Heidelberg
Epon hardener DDSA		Serva, Heidelberg
Epon hardener MNA		Serva, Heidelberg
Formaldehyde		Merck, Darmstadt
Glutaraldehyde 70% (research grade)		Serva, Heidelberg
Potassium dihydrogen phosphate 1/15 *M* solution		Merck, Darmstadt
Cobalt-II-Sulfate	p.a.	Merck, Darmstadt
Maleic acid		Merck, Darmstadt
Metacrylate		Polysciences Inc., Warrington
JB-4 Embedding Comp. A		
JB-4 Solution B		
JB-4 Catalyst		
Horseradish peroxidase (MW 40,000)		Merck, Darmstadt
Sodium hydrogen carbonate	p.a.	Merck, Darmstadt
Osmium tetroxide		Merck, Darmstadt
Sulfuric acid 0.5 *M*		Merck, Darmstadt
Sucrose		Merck, Darmstadt
Trisoxymethylaminomethane		Merck, Darmstadt
Uranyl acetate	p.a.	Merck, Darmstadt

p. a. = per analysis.

3.1.2. Solutions

Ammonium Sulfide Solution

Mix 0.5 ml 20% ammonium sulfide with 99.5 ml distilled water. Add 1–2 drops of 0.2 *M* cobalt sulfate solution to 50 ml of this solution. Filter and shake the filtrate vigorously. Take up the precipitate with the remaining 50 ml of ammonium sulfide solution (stable for 3 h).

Lead Citrate Solution (0.3%)

Pour ca. 80 ml of distilled water over 0.3 g of lead citrate. Add 0.2 ml of 10 *N* NaOH and shake vigorously. Then add a maximum of 0.2–0.4 ml 10 *N* NaOH until the lead citrate is fully dissolved. Carefully cover the solution with paraffin and allow it to stand for 1 day before use (protect against light).

Cacodylate Buffer 0.1 *M*

Dissolve 21.4 g of cacodylic acid sodium salt in 1 liter of distilled water. Adjust with HCl (0.1 *N*) to pH 7.2.

Diamox Solution (10^{-5} *M*)

Open 1 ampule of 250 mg Diamox and add 1 ml of it to 13 ml of distilled water. Add 20 µl of this solution to 57 ml of incubation solution (corresponds to a 10^{-5}*M* solution).

Epon

Mixture A:	62 ml	Epon 812
	100 ml	Hardener DDSA
Mixture B:	100 ml	Epon 812
	99 ml	Hardener MNA

Mixture C: Mix 2 parts A with 3 parts B and add 2% accelerator DMP 30.

Glutaraldehyde Solution

2.5% glutaraldehyde solution in 0.1 *M* cacodylate buffer (pH 7.2).

Incubation Medium (Carboanhydrase Reaction)

1 ml 0.2 *M*	Cobalt sulfate solution
6 ml 0.5 *M*	Sulfuric acid solution
10 ml 1/15 *M*	Potassium hydrogen phosphate solution

– mix –

Just before using, dissolve 0.75 g of sodium hydrogen carbonate in 40 ml distilled water and add. The solution will remain stable for only 10 min.

Incubation Medium (Peroxidase Reaction)

Dissolve 5 mg of 3,3'-diaminobenzidine tetrahydrochloride in 10 ml of 0.05 *M* tris-HCl-buffer pH 7.6 and add 0.1 ml of 1% H_2O_2 (freshly prepared from 30% H_2O_2). The medium is stable for only about 1 h.

Metacrylate
Solution 1: Dissolve 1 g IB-4 catalyst in 100 ml comp. 4.
Solution 2: (Mix directly before use) Mix 1 part solution B with 25 parts solution 1.

Osmium Solution
2% osmium tetroxide solution in 0.1 *M* cacodylate buffer (pH 7.2).

Tris-HCl-Buffer (0.05 *M*)
25 ml of 0.2 *M* trisoxymethylaminomethane (adjust with HCl [0.1 *N*] to pH 7.6). Distilled water ad 100 ml.

Uranyl Acetate Solution
0.5% uranyl acetate solution in 70% ethanol.

Wash Solution (Carbonic Anhydrase Reaction)
9 ml of 1/15*M* potassium dihydrogen phosphate.
1 ml of 1/15 *M* disodium hydrogen phosphate.
9 g of potassium chloride.
Distilled water ad 1,000 ml.

3.2. *Methods*

3.2.1. Transmission Electron Microscopy

- Fix 2×2×2 mm tissue specimens for 3 h in 2.5% glutaraldehyde solution.
- Wash twice for 30 min each time in 0.1 *M* cacodylate buffer.
- Contrast in 1% osmium solution. Dilute the osmium stock solution 1 : 2 with cacodylate buffer solution and use this to cover the tissue specimens for 2.5 h.
- Wash twice for 1 h each in 0.1 *M* cacodylate buffer; block contrast with uranyl acetate, then dehydrate in increasing concentrations of alcohol.

Soak the tissue specimens 0.5 h in 70% alcohol, then allow to stand over night in the 0.5% uranyl acetate solution.
Then cover the specimens 0.5 h each with 80% ethanol.
Then cover the specimens 0.5 h each with 90% ethanol.
Then cover the specimens 0.5 h each with 96% ethanol.
Then cover the specimens 0.5 h each with 100% ethanol.
Then cover the specimens 0.5 h each with 2 parts 100% ethanol and 1 part acetone.
Then cover the specimens 0.5 h each with 1 part 100% ethanol + 2 parts acetone.
Then cover the specimens 0.5 h each with acetone.
Allow to stand over night in a mixture of 1 part epon (mixture C) and 1 part acetone.

- Allow the acetone to drain for about 4 h.
- Fill forms with epon mixture C, embed the tissue.
- Allow the epon to polymerize 24 h at 60 °C and 48 h at 90 °C.
- Store the specimens for 1–2 weeks until completely hardened.
- Section contrasting:
  15 min 0.5% uranyl acetate solution (10 min illuminated, 5 min unilluminated).

Rinse 3× with distilled water.
5 min 0.3% lead citrate solution.
Rinse 3× with distilled water.

3.2.2. Scanning Electron Microscopy

– Fixing, refixing and dehydration as in transmission electron microscopy (omit block contrasting with uranyl acetate).
– Place the tissues in 100% acetone solution into the critical-point-dryer (Balzers Company, Wiesbaden) and dry.
– After drying, glue the preparations with conducting silver (Leitsilber[200], Demetron Co., Hanau) to an SEM holder (Balzers Co., Wiesbaden).

3.2.3. Carbonic Anhydrase Reaction [69, 108]

– Fix 2×2×4 mm tissue specimens in 2.5% glutaraldehyde solution for 2–4 h, then wash twice with cacodylate buffer. Store for no more than 4 days in cacodylate buffer.
– Immerse the tissue for 15 min in 50% water-metacrylate solution 1.
– Immerse the tissue for 15 min in 70% water-metacrylate solution 1.
– Immerse the tissue for 35 min in 100% water-metacrylate solution 1.
– Immerse the tissue for 35 min in 100% water-metacrylate solution 2.

Polymerization follows within 20 min in the exsiccator. Refrigerate the embedded tissue (in the exsiccator) until needed.

– Prepare 2-µm thick sections and collect them in a Petri dish with aqua dest. They must float on the surface. Removal with a glass spatula.
– Allow the sections to float 1–9 min in incubation medium in a Petri dish.
– Allow the sections to float for 1 min in wash solution.
– Allow the sections to float for 3 min in ammonium sulfide solution.
– Allow the sections to float for 1 min in distilled water.
– Allow the sections to float for 1 min in distilled water.
– Allow the sections to float for 1 min in distilled water.

Negative check: Incubate the floating sections 1–9 min in a 10^{-5} *M* Diamox solution instead of incubation medium; all other steps as above.

– Allow the sections to dry on a glass slide and mount them. Use the preparations within a week.

3.2.4. Peroxidase Reaction [92]

– Fix the tissue slices (3–4 mm thick) in an aldehyde-buffer mixture (pH 7.2 final concentrations: 2% formaldehyde, 2.5% glutaraldehyde, 0.25% $CaCl_2$, 0.1 *M* cacodylate buffer) for 5 h at 20 °C.
– Wash out over night in cacodylate buffer pH 7.2.
– Prepare frozen sections 40 µm thick.
– Place the sections for 3–10 min in a Petri dish with incubation medium (20 °C).
– Rinse the sections 3× with distilled water.
– Re-fix the sections in 1.3% osmium solution pH 7.2 with 5% sucrose for 90 min.
– Dehydrate, embed in epon; section contrasting with lead citrate solution (0.3%).

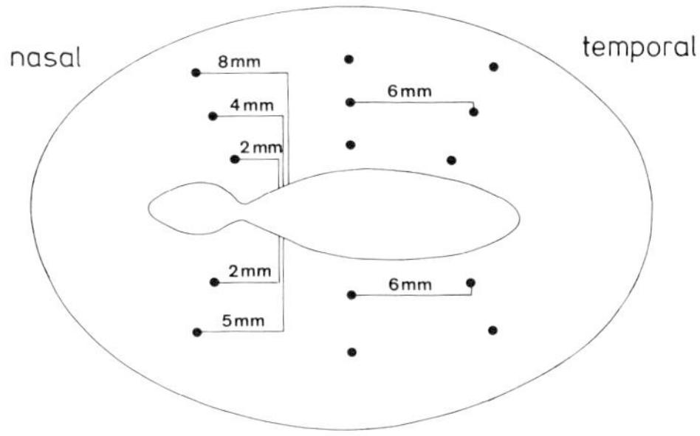

Fig. 3. Location of the 15 quantitatively evaluated regions of albino rabbits and cynomolgus monkeys.

3.3. Animal Conjunctiva

3.3.1. Albino Rabbits

3.3.1.1. Ultrastructure of Rabbit Conjunctiva

The conjunctiva of both eyes of a total of four albino rabbits (New Zealand, ca. 12 months old) was studied. The animals were killed with 5–10 ml of a saturated KCl solution, and the upper and lower lids removed and stretched out on a cork tray. Tissue specimens 2–3 mm^2 in size were then removed from nine specific regions of the upper lid and six specific regions of the lower lid (fig. 3).

The specimens were prepared for TEM and SEM. Semithin slices were stained for orientation with the method of Richardson et al. [152]. Ultrathin sections were prepared with the Ultracut (Reichert Co., Heidelberg). Following section contrasting with 0.3% lead citrate, the sections were studied with the Siemens Elmiscope 101A electron microscope. Using photomontages (magnification: × 3,000), at least 50 epithelial surface cells were evaluated in each region on the basis of their organelle content. Each of these epithelial cells was in direct contact with the surface of the epithelium, and it was necessary to section their cell nuclei as well (see 4.1.1.1. for further criteria of evaluation). The tissue specimens prepared for SEM were studied with a Leitz Co. AM 1000 SEM (Wetzlar).

3.3.1.2. Resorption Processes in Rabbit Conjunctival Epithelium

Three albino rabbits (New Zealand, ca. 12 months old) were used for this study. Horseradish peroxidase (HP) was inserted into the conjunctival sac in a concentration of 5 mg/ml (physiologic salt solution) and the lids were closed. After incubation periods of 5, 30 and 60 min, respectively, the animals were killed with a saturated KCl solution. After excision of the

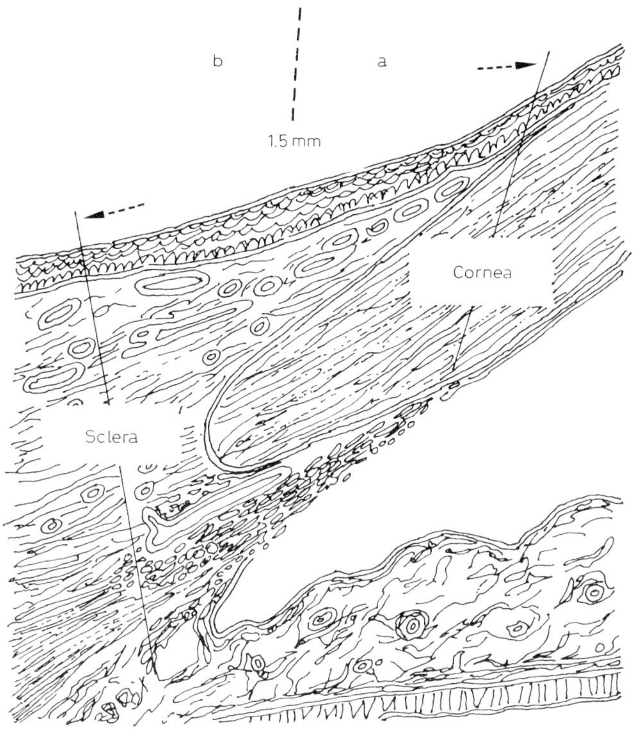

Fig. 4. Diagram of ocular limbus. *a*, *b* sections near to and far from the limbus epithelium respectively.

upper and lower lids, we removed conjunctival specimens 2–3 mm² in size. The tissue specimens were then stained for peroxidase with the Karnovsky method [92]. After embedding them in epoxy, we made ultrathin sections. Uncontrasted and contrasted sections were then assessed with a Siemens Elmiscope 101A electron microsope.

3.3.1.3. Corneal Reepithelialization at the Rabbit Limbus: Ultrastructure and Histochemistry

Four additional tissue specimens were taken at the limbus of each of the 4 albino rabbits (see 3.3.1.1.). The semitransparent area of the peripheral cornea where the episcleral capillaries begin (fig. 4) was defined as the border between cornea and limbus. Triangular pieces of cornea with a scleroconjunctival strip some 2 mm wide were excised at 3, 6, 9 and 12 o'clock. After embedding of the specimens in epoxy, two limbus zones a (near the cornea) and b (distant from the cornea) were distinguished with the semithin slices (fig. 4). With the aid of ultrathin sections, the epithelial surface cells of this total of eight regions were evaluated by the method described earlier. Some of the tissue specimens were embedded in metacrylate for carbonic anhydrase determination of the conjunctival epithelium by the Hansson method [69].

Fourteen albino rabbits (New Zealand, ca. 6 months old) were used for more accurate determination of chronological and morphological aspects of the metaplastic transformation of limbal to corneal epithelium during the cornea's reepithelialization phase. Under local anesthesia (Benoxinat® 0.4%/[Oxybuprocaine]) and general sedation (5 mg Valium® [Diazepam] i.v.), the corneal epithelium was abraded with a Q-tip. The abrasion extended to the corneolimbal border, as mentioned, so that the limbal epithelium remained circularly intact (fig. 4).

One rabbit developed progressive corneal opacification within 48 h after abrasion. Another exhibited progressive corneal vascularization after 4 days. These 2 animals were not included in the study; each was replaced by another rabbit.

Two rabbits each were killed at intervals of 12, 24 and 48 h and after 4, 7 and 12 days with a saturated KCl solution. Two rabbits served as controls. The cornea, including a 2 mm wide scleroconjunctival strip, was removed and divided into six triangular segments. The carbonic anhydrase reaction test was performed on each of two tissue specimens from each eye, using the method described in 3.2.3. above. All positive reactions (described later in Results) proved suppressible with Diamox, a specific inhibitor of carbonic anhydrase.

Two further tissue specimens were embedded for TEM. Ultrathin sections were prepared from the epithelial cell with farthest forward proliferation onto the cornea. For orientation, we used semithin sections stained with the Richardson method [152]. The ultrathin sections were then studied with a Zeiss (Oberkochen) EM 10 electron microscope. The SEM study of the remaining specimens was done with the Leitz (Wetzlar) AM 1000 SEM.

3.3.2. *Cynomolgus Monkeys*

3.3.2.1. Ultrastructure of Monkey Conjunctiva

The upper and lower lid conjunctiva of both eyes of a total of three cynomolgus monkeys was studied. The monkeys were killed with a

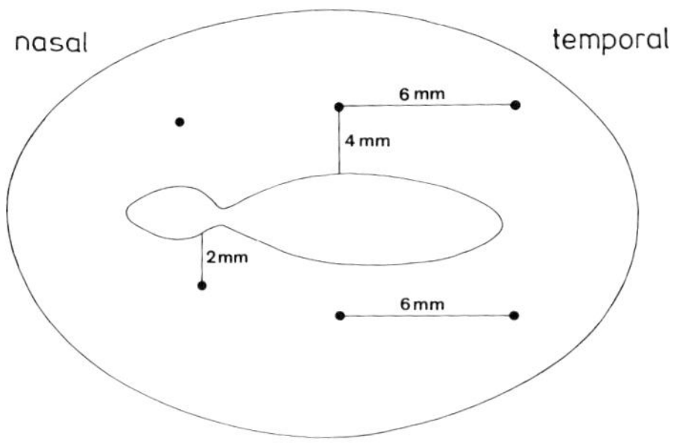

Fig. 5. Excision regions of conjunctival specimens from cynomolgus monkeys after pilocarpine application.

saturated KCl solution. The excision regions and the method of preparation were identical to those described in section 3.3.1.1. above, as were the criteria of evaluation.

3.3.2.2. Resorption Processes in Monkey Conjunctival Epithelium

Three cynomolgus monkeys were used for this part of the study. The method was identical to that of the rabbit experiments (see 3.3.1.2.).

3.3.2.3. Effect of an Antiglaucomatous Agent (Pilocarpine) on the Conjunctival Ultrastructure of Cynomolgus Monkeys (Two Application Forms)

A rigid medicine dispenser (plastic ring, ca. 13 mm in diameter) which constantly released an unchanging amount of pilocarpine (ca. 30 ng/h) was placed in the conjunctival sac of each eye of two cynomolgus monkeys for 2 months. This rigid form was chosen because commercially available medicine containers (e.g. Ocusert) were invariably lost under experimental conditions. The rigid dispensers caused mechanical irritation of the conjunctiva with each blink of the lids. They were exchanged every 3 days. In addition, each right eye of 2 cynomolgus monkeys was given drops of a watery 4% pilocarpine solution 3 times a day (ca. 6 mg/day) for 3 months. The untreated left eye served as a control.

The animals were then killed with a saturated KCl solution and specimens of conjunctival epithelium removed bilaterally as described above. The locations of removal are shown in figure 5. The tissue samples were fixed for electron microscopy as described and embedded in epoxy. We analyzed the ultrathin slices with a Zeiss (Oberkochen) EM 10 electron microscope, using the evaluation criteria of epithelial surface cells described in 3.3.1.1.

3.4. *Human Conjunctiva*

3.4.1. *Healthy Conjunctival Tissue*

Tissue specimens were removed from a total of 48 clinic patients with no detectable conjunctival disease. The removal was performed under topical anesthesia with Chibro-Kerakain® (Proxymetacaine) eye drops or during surgery under retrobulbar or general anesthesia. One specimen was taken from the limbus area at 12 o'clock, six from the bulbar conjunctiva, six from the conjunctival fornix and six from the upper edge of the tarsus of the upper and lower lid (see fig. 6 for a diagram of the removal locations from human conjunctiva). The specimens were excised according to a standardized procedure. The patients were classified into four age groups (group 1, 1–20 years; group 2, 21–40 years; group 3, 41–60 years; group 4, 61 years and older). Patient data (age, removal location, reason for clinic stay) are given in table 2. The excised tissue specimens were processed immediately as described in 3.2.1. The same criteria were used for the electron microscopical evaluation as in 3.3.1.1. However, the evaluation took place directly at the electron microscope (Zeiss [Oberkochen] EM 109) under enlargements between ×3,000 and ×20,000. Representative findings were recorded photographically.

In each age group, the mean ($\bar{x}$) and the standard deviation (s) of each cell type were calculated for each region. The following formula was used to calculate the standard deviation:

$$S_x = \sqrt{\frac{\Sigma x - \frac{(\Sigma x)^2}{n}}{n-1}}$$

x ≙ individual measurements; n ≙ total number of measurements.

These figures were also obtained for the individual epithelial regions independent of patient age.

3.4.2. *Pathologically Altered Conjunctival Tissue*

In all, three of the regions described earlier were chosen for evaluation of pathological tissue specimens. Under topical anesthesia (Chibro-Kerakain® [Proxymetacaine]), the specimens were removed from the superior temporal bulbar conjunctiva, the superior medial fornix area, and the inferior medial tarsal conjunctiva (fig. 6). The mean values of the random distribution of cell types in specimens from pathologically altered conjunc-

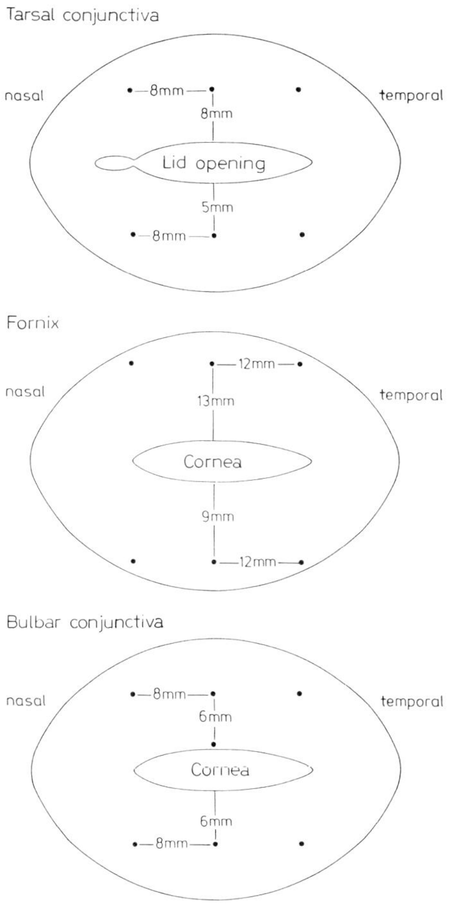

Fig. 6. Diagrams of removal locations of human conjunctiva.

tiva ($\bar{x}_1$) were compared statistically with those of random cell types from normal subjects ($\bar{x}_2$), independent of age. Inasmuch as the random specimens of both groups had no normal distribution, a transformation with the arc-sine function was used. With this data, the t-test was then applied using the following formula:

$$t = \frac{\bar{x}_1 - \bar{x}_2}{s} \cdot \sqrt{\frac{n_1 \cdot n_2}{n_1 + n_2}}$$

$$s = \sqrt{\frac{(n_1-1)s_1^2 + (n_2-1)s_2^2}{n_1 + n_2 - 2}}$$

Table 2. Data of the 48 healthy subjects

Patient	Born	Eye	Date of excision	Reason for clinic stay
Bulbar conjunctiva				
Age 1–20				
E.M.	4/24/77	L	7/9/81	Penetrating KP
M.E.	1/29/66	R	8/13/81	Secondary cataract
N.Sch.	9/2/77	R	2/12/82	Squint OP
Y.K.	7/15/74	L	1/28/82	Squint OP
21–40 years				
C.Sch.	9/8/56	L	3/30/81	Squint OP
B.B.	5/24/51	L	3/31/81	Penetrating KP
J.M.	7/13/50	R	2/8/82	Squint OP
M.H.	3/15/45	L	2/22/82	Squint OP
41–60 years				
F.B.	8/13/24	R	8/18/81	Cat. Ex.
H.S.	2/14/40	R	4/22/81	Penetrating KP
A.G.	10/10/28	L	7/8/85	Cat. Ex.
G.G.	6/7/39	L	7/9/85	Cat. Ex.
↑61 years				
C.Sch.	1/18/98	R	4/30/81	Cat. Ex.
W.P.	6/17/99	R	4/2/81	Cat. Ex.
E.G.	12/3/01	L	4/24/81	Cat. Ex.
A.Sch.	1/29/00	R	4/1/81	Cat. Ex.
Conjunctival fornix				
Age 1–20				
J.B.	10/17/81	L	6/19/85	Squint OP
M.Sch.	7/12/82	R	2/18/83	L: buphthalmus
C.B.	6/20/82	L	7/12/83	Cat. Ex.
F.Sch.	8/15/65	L	12/30/82	Perf. cornea-iris-lens injury
Age 21–40 years				
R.M.	5/31/48	L	7/31/85	Cat. Ex.
E.W.	4/26/64	R	4/14/86	Cat. Ex.
N.L.	10/22/56	L	7/9/85	Perf. cornea-lens injury
G.Sch.	3/6/61	L	7/23/85	Squint OP

Table 2 (continued)

Patient	Born	Eye	Date of excision	Reason for clinic stay
Age 41–60 years				
R. H.	10/24/37	R	7/15/85	Cat. Ex.
P. H.	10/23/44	L	6/20/86	Squint OP
W. Sch.	11/6/38	R	7/26/83	Cat. Ex.
G. D.	7/8/30	L	2/17/83	Cat. Ex.
↑61 years				
L. P.	6/7/22	L	8/1/83	Cat. Ex.
G. D.	10/31/08	R	2/17/83	Cat. Ex.
A. F.	12/7/03	R	8/1/83	Cat. Ex.
M. B.	4/28/20	R	7/26/83	Cat. Ex.
Tarsal conjunctiva				
Age 1–20				
Th. R.	1/14/80	R	7/12/85	Squint OP
S. N.	5/26/79	L	7/19/85	Squint OP
S. W.	7/8/76	R	7/22/85	Squint OP
N. E.	3/13/78	L	7/19/85	Squint OP
Age 21–40 years				
A. T.	12/29/56	R	7/10/85	Squint OP
R. M.	5/31/48	L	7/31/85	Cat. Ex.
H.-P. P.	1/19/51	L	7/17/85	Cat. Ex.
K. B.	5/1/50	L	7/17/85	Cat. Ex.
Age 41–60 years				
J. L.	9/8/32	L	8/20/85	Cat. Ex.
R. V.	12/22/28	R	8/23/85	Cat. Ex.
G. B.	4/23/37	L	7/18/85	Cat. Ex.
H. Sch.	12/15/31	R	7/12/85	Cat. Ex.
↑61 years				
E. H.	2/4/21	R	8/29/85	Cat. Ex.
I. H.	6/19/07	R	8/19/85	Cat. Ex.
E. Sch.	9/29/99	R	8/27/85	Cat. Ex.
E. E.	3/10/08	R	8/21/85	Cat. Ex.

The individual steps in calculation were performed with a specially programmed Atari computer. Differences in mean values of random specimens were termed significant when the probability of error was 1% and highly significant when it was 0.1%.

3.4.2.1. The Conjunctiva in Dry Eye (Sjögren's Syndrome; table 3)

The conjunctiva of 6 patients with internistically confirmed Sjögren's syndrome was studied. To determine the volume of tear secretion, a standardized strip of filter paper (Dr. Mann Co., Berlin) was inserted into the conjunctival sac under local anesthesia (Chibro-Kerakain® [Proxymetacaine]) after absorption of the lacrimal fluid already present in the lateral

Table 3. Data of patients with Sjögren's syndrome (SS)

Patient		Born	Eye	Excision date	Date SS diagnosed	Schirmer I
G.R.	♀	2/6/33	L	4/10/86	10/85	R 1 mm L 1 mm
L.K.	♀	9/18/24	L	4/9/86	7/81	R ∅ L 1 mm
E.L.	♀	9/20/41	L	3/13/86	12/85	R ∅ L ∅
J.F.	♀	3/20/30	R	3/11/86	11/85	R ∅ L ∅
K.H.	♀	12/19/36	R	10/3/86	2/86	R 2 mm L 1 mm
S.V.	♂	1/23/25	R	3/10/86	5/80	R 1 mm L ∅

Table 4. Data of patients with lime burn

Patient	Born	Eye	Date of burn	Burn degree	Date of excision
M.W.	1/4/64	R	8/27/80	III	4/2/86
M.G.	7/29/46	L	6/1/85	III	3/14/86
T.G.	12/29/66	L	6/24/85	III	3/11/86
A.B.	1/12/37	R	8/20/83	IV	3/12/86
F.W.	10/8/27	R	7/24/56	III	3/11/86
K.R.	3/19/23	L	5/10/75	IV	3/6/86

lower lid area. The examination took place at medium room illumination. The saturated band on the paper strip was measured after 5 min.

3.4.2.2. The Conjunctiva after Lime Burn (table 4)

The patients with lime burn showed all of Reim's chronic characteristics of third- to fourth-degree lime burn [149].

3.4.2.3. Effect of Antiglaucomatous Agents (Pilocarpine, Timolol) on Conjunctival Ultrastructure

Six patients were selected for each group; all had had antiglaucoma monotherapy for at least one year with either timolol (Chibro-Timoptol®) 0.5% or at least 1% pilocarpine (Isopto-Pilocarpin®, Pilopos®, Spersacarpin®) containing benzalconium chloride as preservative. Patient data are given in tables 5 and 6.

Table 5. Patient data upon conjunctival tissue excision after pilocarpine monotherapy (at least 1%)

Patient	Born	Eye	Tissue removed	Duration of monotherapy years
S.W.	4/6/31	R	4/24/86	8
S.J.	10/28/05	R	4/4/86	12
H.K.	11/19/22	L	4/4/86	2
B.L.	1/3/15	L	3/18/86	9
E.M.	1/13/05	R	3/24/86	5
J.R.	12/19/11	R	10/3/86	4

Table 6. Patient data upon conjunctival tissue excision after timolol monotherapy (0.5%)

Patient	Born	Eye	Tissue removed	Duration of monotherapy years
H.W.	5/4/23	L	6/10/86	3
J.K.	1/7/21	R	5/30/86	2
K.P.	12/7/13	R	5/9/86	1
E.E.	12/13/20	R	5/9/86	3
L.L.	5/27/23	L	4/23/86	9
G.B.	7/14/13	R	4/7/86	4

4. Results

4.1. Animal Conjunctiva

4.1.1. Ultrastructure of Conjunctival Epithelium in Albino Rabbits and Cynomolgus Monkeys

Goblet cells were already discernible from other epithelial cells in the SEM view of the conjunctiva (fig. 7); the TEM overview showed both goblet cells and dark and light-colored cell types, depending on the organelles and the glycogen content of the epithelial cells (fig. 8). However, no conformity to a pattern of differentiation was detectable in the epithelial surface cells on the basis of these characteristics.

This became possible only after we assigned certain cell organelles to certain cell types and studied the latter's distribution in various regions of the conjunctiva. We defined 5 different cell types on the basis of ultrastructural criteria. The definitions were based both on the number and kind of organelles found in the cells and on the arrangement of these organelles in the cytoplasm.

4.1.1.1. Definition of the Five Cell Types

Type I Cells. We designated the already familiar goblet cells as type I cells. They are easily recognized by their large, electron-dense granules, which more or less fill the cytoplasm (fig. 9).

In most cases, a well-developed Golgi system is found in the perinuclear space of cells of this type. Depending on secretion pressure, the cell organelles and the cell nucleus of a fully developed goblet cell are pushed into the marginal – especially the basal – cell regions by the numerous mucous enclosures. The final stage of this cycle of secretion is characterized by an empty goblet cell which has released its secretory granules and other cell organelles to the epithelial surface.

Type II Cells. Type II cells of cynomolgus monkeys were defined as those with numerous 60–300 nm large electron-dense granules (fig. 10).

a b

Fig. 7. a, b SEM views of the conjunctival epithelium of a cynomolgus monkey. Note the polygonal cell pattern, the stomata of the goblet cells (arrow), the opening to a conjunctival crypt (arrowheads) and the numerous microvilli on the cell surface. *a* × 423; *b* × 576.

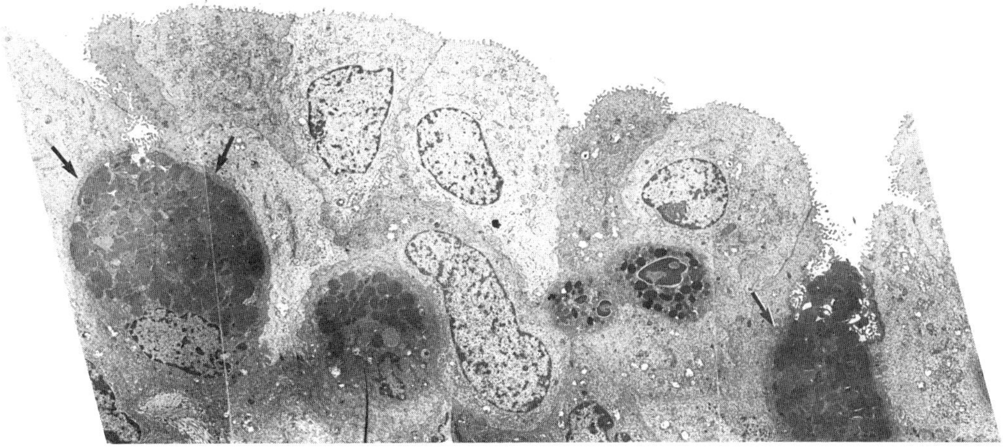

Fig. 8. Electron micrograph of the conjunctival epithelium of an albino rabbit, showing both goblet cells (arrows) and dark and light-colored epithelial cells. × 2,290.

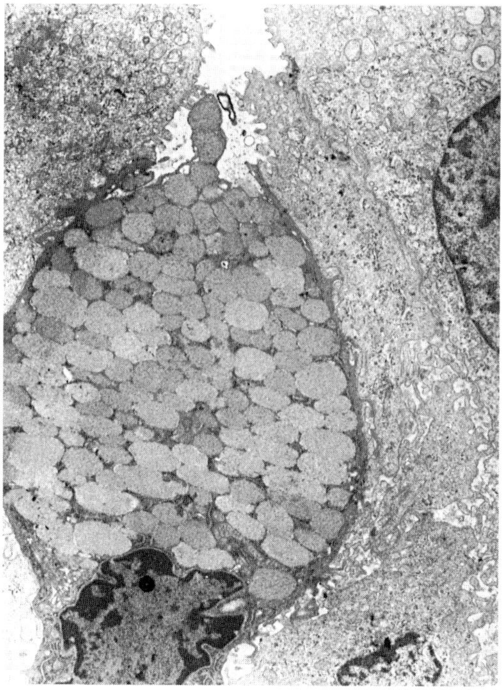

9

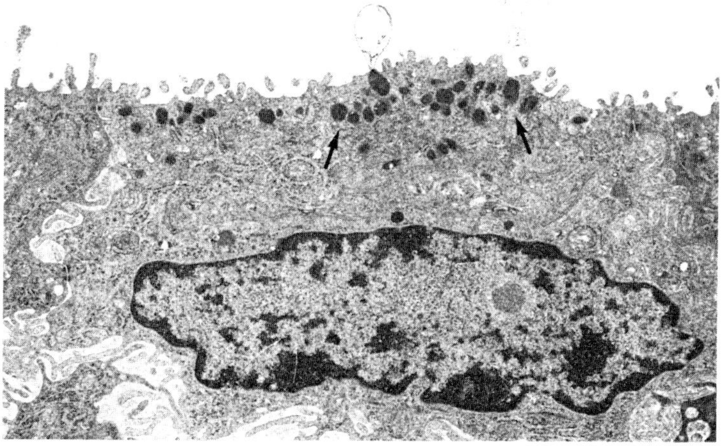

10

Fig. 9. Electron micrograph of a type I (goblet) cell of rabbit conjunctiva. × 6,500.

Fig. 10. Apical area of a type II cell of the conjunctival epithelium of a cynomolgus monkey. Note the numerous electron-dense granules (arrows). × 8,750.

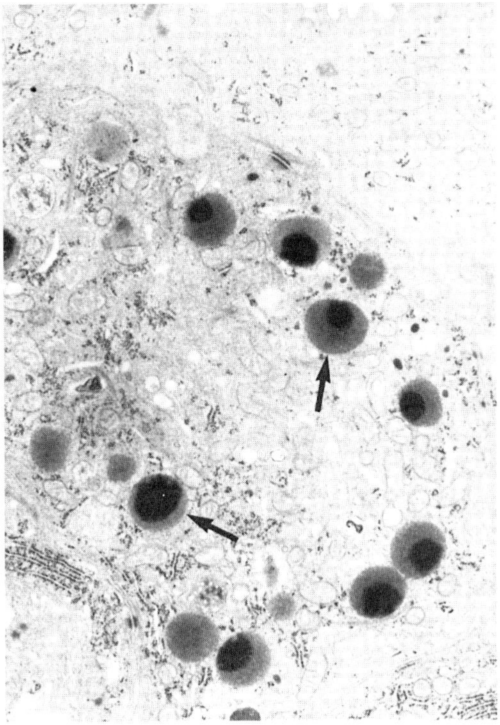

Fig. 11. Electron micrograph of a type II cell of rabbit conjunctiva with typical osmiophilic granules (arrows). × 10,500.

The granules are found for the most part in the apical cytoplasm of the cells. These granules are usually more electron-dense toward the surface of the epithelium. They are surrounded by a double membrane and often come into contact with the apical cell membrane. Type II cells also contain rER and Golgi material; however, these organelles are not the dominant structures of this cell form.

The type II cells of rabbits contain very large (200–800 nm) granules which, in contrast to those of the monkey conjunctiva, have a highly osmiophilic core with a rather weakly osmiophilic circumference (fig. 11). Here too, granules are localized in the apical cytoplasm of the epithelial cells. Some mitochondria also appear in these cells, with scattered Golgi complexes and some rER, mostly near the nucleus.

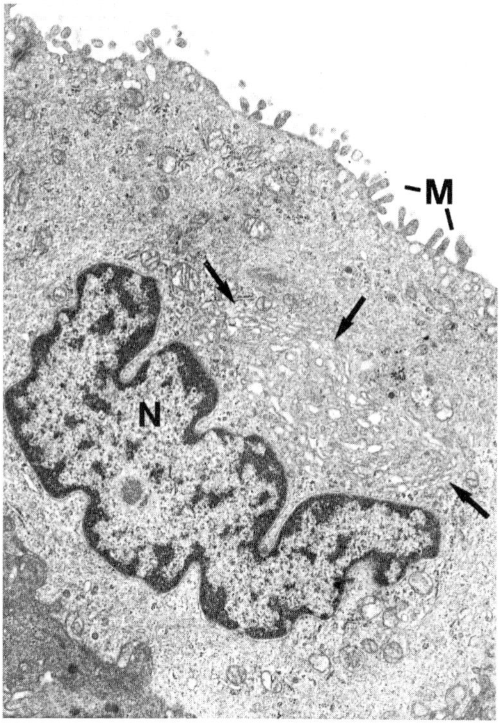

Fig. 12. Electron micrograph of a type III cell. Arrows = Golgi complex; N = cell nucleus; M = microvilli. × 9,200.

Type III Cells. This cell type is recognizable by its well-developed Golgi complex. Several Golgi complexes often lie around the cell nucleus, nearly filling the cytoplasm in the direction of the epithelial surface (fig. 12).

The Golgi complexes consist of up to 15 densely concentrated spaces; these are small centrally but usually somewhat broader in the periphery, and are surrounded by taut or lightly bulging membranes. Numerous vesicles often collect on the concave or convex side of the Golgi complexes. Other cell organelles such as rER and mitochondria are less often present. No essential structural differences are found in this cell type in monkeys and rabbits.

Type IV cells are characterized by large quantities of rER. Membranes with thick coatings of ribosomes are usually found near the nucleus; these

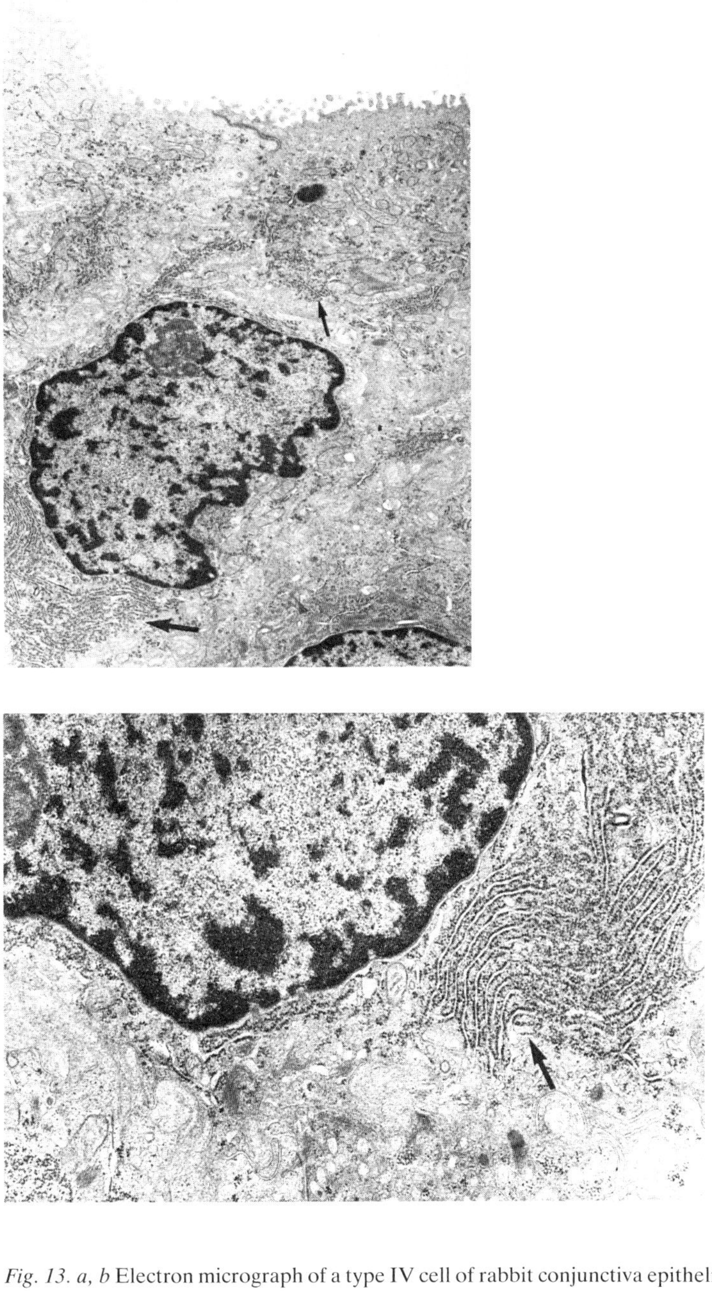

a

b

Fig. 13. a, b Electron micrograph of a type IV cell of rabbit conjunctiva epithelium. Note the well-developed rER (arrows). *b* Enlarged section of *a*. *a* × 7,600; *b* × 15,300.

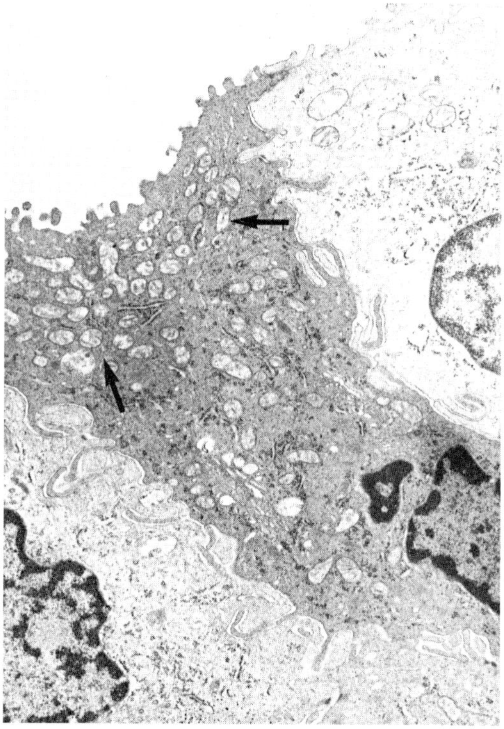

Fig. 14. Electron micrograph of a type V cell. The apical cytoplasm of the epithelial cell is filled by numerous mitochondria (arrows). × 10,500.

form tightly layered cisterns (fig. 13). The cisterns of rER are quite variable in shape and often up to 50 nm wide. Mitochondria and Golgi apparatuses are present in these cells, but do not determine the cellular structure.

Type V Cells. This cell type is characterized by a high content of up to 40 mitochondria in a single cell cross section, typically located in the apical part of the cell. For this reason other cell organelles such as Golgi complexes or rER are rare. The mitochondria have a well-developed system of cristae (fig. 14). They vary in size but are usually up to 1,500 nm long and up to 500 nm wide. The cytoplasm of type V cells is often more electron-dense than that of neighboring epithelial cells.

4.1.1.2. Distribution of Five Cell Types in Rabbit and Monkey Conjunctiva

The quantitative evaluation of the ultrathin sections showed a relatively constant distribution of the 5 different cell types in the regions studied in cynomolgus monkeys and albino rabbits. The percentage of individual cell types among the epithelial surface cells of monkey conjunctiva (a) and rabbit conjunctiva (b) is shown in figures 15–19.

Type I Cells. In monkey conjunctiva, this cell type was found foremost in the lower lid, where 25% of epithelial surface cells belong on the average to cell type I (goblet cells). Goblet cells were also quite numerous nasally and temporally in the upper lid, with ca. 20%. This cell type was rarest in the middle regions of the upper lid. Type I cells were scarcer in the rabbit conjunctiva than in the monkey conjunctiva. In rabbits their frequency diminished from nasal to temporal in both the upper and the lower lid. Type I cells were especially infrequent in the temporal lower lid area.

Type II Cells. In monkey conjunctiva, this cell type (cells with osmiophilic secretory granules) appeared most often in the nasal lid conjunctiva and the upper and lower lids. It was the cell type found most infrequently among surface epithelial cells of the temporal upper and middle lower lids. In epithelial regions in which the second cell type was frequent, it comprised ca. 20% of the epithelial surface cells. Its percentage in other areas, however, was below 10%. On the whole, type II cells were infrequent in rabbits; only in the caudal regions of the nasal and middle upper lid were they quite frequent, with 22 and 16%, respectively, and likewise comprised nearly 20% of epithelial surface cells in the temporal inferior section of the lower lid. Elsewhere their frequency was below 5%.

Type III Cells. Cells with a well-developed Golgi apparatus were least frequent in the monkey and rabbit conjunctiva. In monkey conjunctiva, this cell type appeared relatively frequently only in the temporal upper lid area. In other epithelial regions it constituted barely more than 5% of epithelial surface cells. In rabbit conjunctiva, this cell type appeared with relative frequency only in the temporal upper and lower lid areas. Its frequency in other regions was less than 5%, and occasionally it was not found at all.

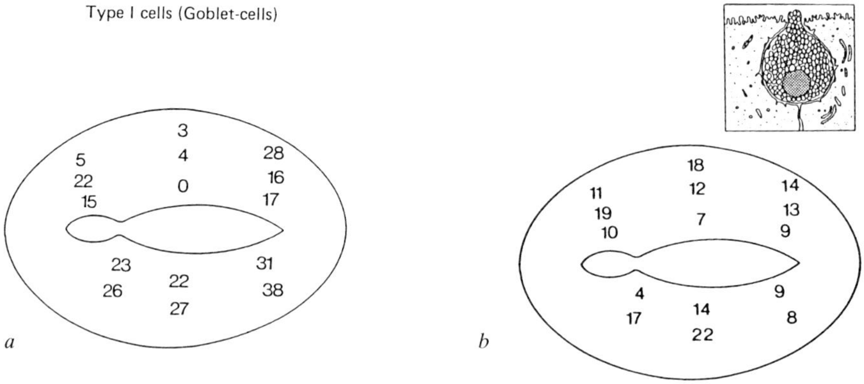
Type I cells (Goblet-cells)
3
4
0
5
22
15
28
16
17
23
26
22
27
31
38
a
18
12
7
11
19
10
14
13
9
4
17
14
22
9
8
b

15

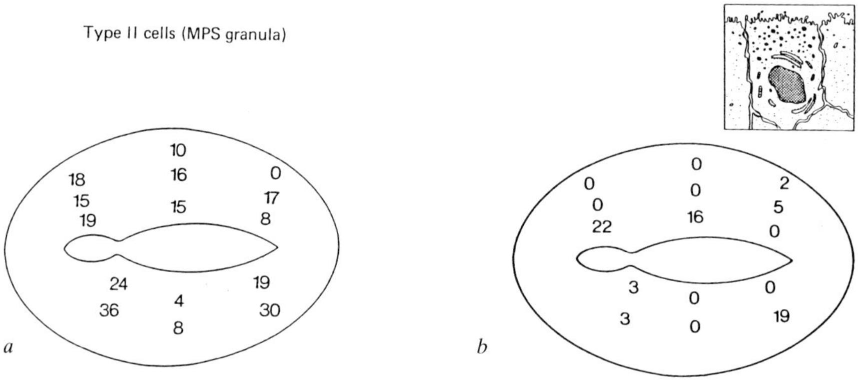
Type II cells (MPS granula)
10
16
15
18
15
19
0
17
8
24
36
4
8
19
30
a
0
0
16
0
0
22
2
5
0
3
3
0
0
0
19
b

16

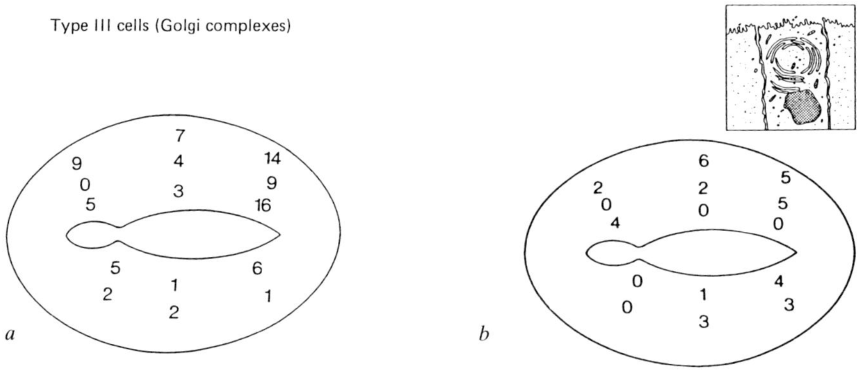
Type III cells (Golgi complexes)
7
4
3
9
0
5
14
9
16
5
2
1
2
6
1
a
6
2
0
2
0
4
5
5
0
0
0
1
3
4
3
b

17

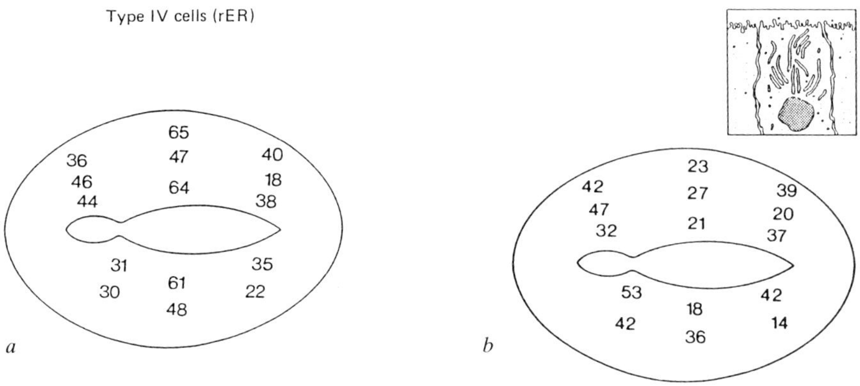

18

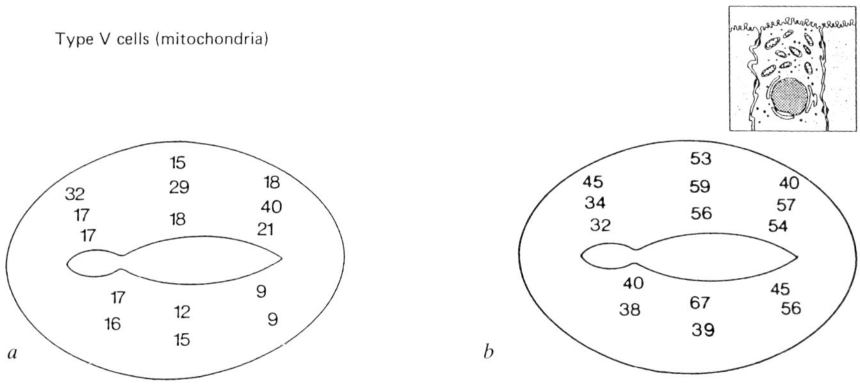

19

Fig. 15. Average percentage of type I cells among epithelial surface cells. *a* Monkey conjunctiva. *b* Rabbit conjunctiva.

Fig. 16. Percentage of type II cells among epithelial surface cells. *a* Monkey conjunctiva. *b* Rabbit conjunctiva.

Fig. 17. Percentage of type III cells among epithelial surface cells. *a* Monkey conjunctiva. *b* Rabbit conjunctiva.

Fig. 18. Percentage of type IV cells among epithelial surface cells. *a* Monkey conjunctiva. *b* Rabbit conjunctiva.

Fig. 19. Percentage of type V cells among epithelial surface cells. *a* Monkey conjunctiva. *b* Rabbit conjunctiva.

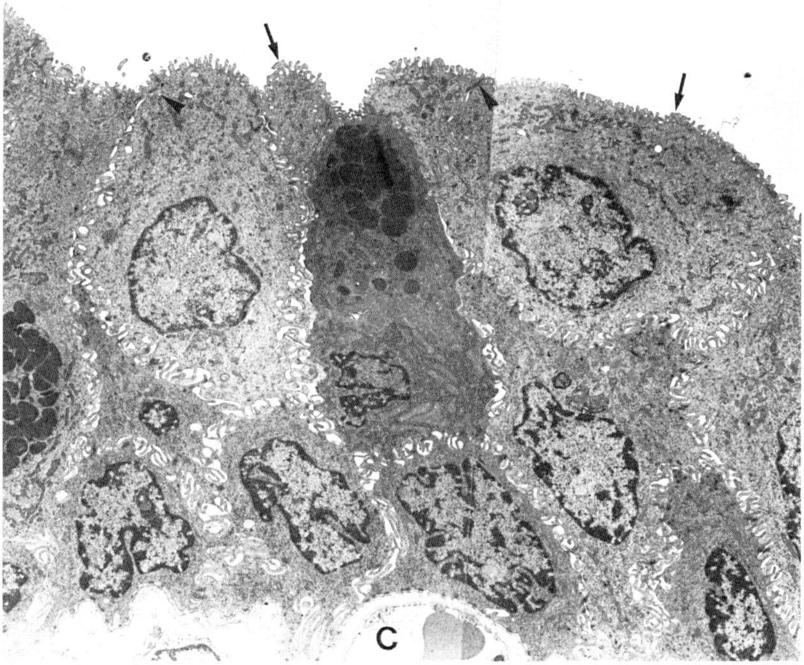

Fig. 20. Electron microscopic overview of rabbit conjunctiva. Note the many microplicae (arrows), the tight junctions (arrowheads), and the fenestrated capillaries (C) in the subepithelial conjunctival tissue. BM = Basement membrane. × 3,800.

Type IV Cells. In monkey conjunctiva, ergastoplasma-rich epithelial cells were the most frequent of all, constituting up to two thirds of epithelial surface cells. They were concentrated most in the middle part of the upper and lower lids. They gradually decreased in number temporally and nasally. They were invariably found more often near the lid edges than in the conjunctiva regions near the fornix. In rabbit conjunctiva, this cell type was particularly frequent in the nasal sectors of the upper and lower lids, where they comprised on the average more than 40% of all cells. They constituted some 30% of epithelial surface cells in the middle and temporal lid sections; this figure sank as low as 14% in the temporal lower lid area. Type IV cells were the second most frequent ones in rabbit conjunctiva.

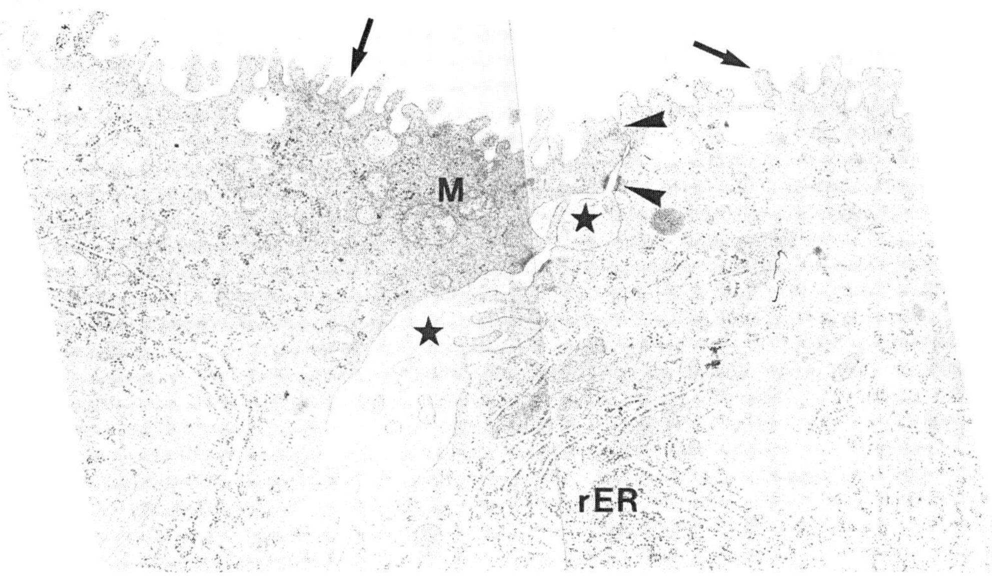

Fig. 21. Electron micrograph of the apical section of 2 surface epithelial cells of the conjunctiva of a cynomolgus monkey. Note the highly expandable intercellular spaces with many finger-shaped cellular extensions (stars). Arrows = Microvilli; arrowheads = tight junction; rER = rough endoplasmatic reticulum; M = mitochondria. × 19,500.

Type V Cells. In monkeys, mitochondria-rich cells were chiefly present in the temporal upper lid area; their numbers were lowest in the temporal lower lid area, with ca. 10%. In monkey conjunctiva they were second highest in frequency. In rabbit conjunctiva, type V cells were the most frequent of all five cell types. This cell type was especially represented in the middle and temporal sections of the upper lid and in the temporal area of the lower lid, with more than 50% of the epithelial surface cells.

4.1.2. Resorption Processes in Rabbit and Monkey Conjunctival Epithelium

4.1.2.1. Morphological Basis of Resorption Processes in the Conjunctival Epithelium

The SEM view (fig. 7) shows the quantity of microvilli covering the surface of the conjunctival epithelial cells. The TEM micrographs show that neighboring microvilli often coalesce to form T- or L-shaped microplicae (fig. 20).

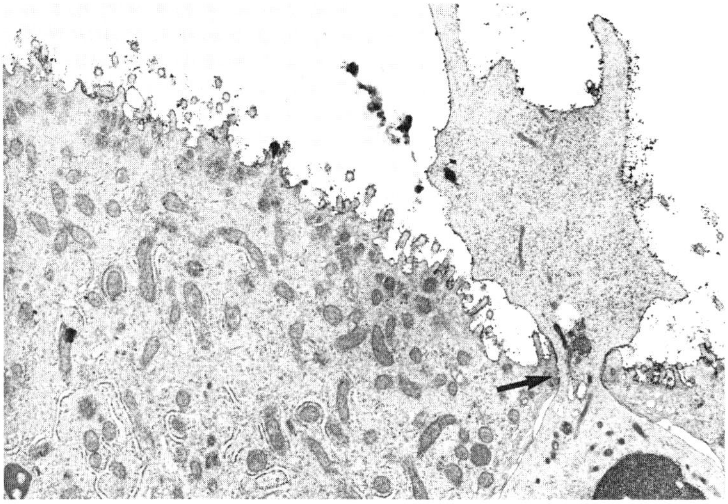

Fig. 22. The apical portion of a type V cell of monkey conjunctiva. Note the cell which is migrating through the intercellular spaces (arrow). × 8,300.

The epithelial cells were firmly bound together toward the surface by tight junctions. The bond was further strengthened by additional desmosomes (fig. 21).

In this way the intercellular spaces remained tightly closed toward the surface. The basal portions of the intercellular spaces, on the contrary, were highly expandable and contained small, interlocking cellular extensions (fig. 21). Cells often migrated through these spaces (fig. 22).

Capillaries extending directly to the basal lamina were often present in the subepithelial conjunctival tissue. The endothelium of these capillaries was clearly fenestrated (fig. 20).

4.1.2.2. HP Resorption in Monkey Conjunctiva

When monkey conjunctiva was incubated for 5 min with HP, the epithelial surface became coated with a 0.15-μm thick layer of reaction product. In addition, numerous invaginations and pinocytotic vesicles filled with reaction product were found in the adjacent cytoplasm (fig. 23). Reaction product was not yet present in the intercellular spaces at this point.

After 30 min of incubation, pinocytotic vesicles were already visible in the deeper parts of the surface epithelial cells. Most of them, however, were

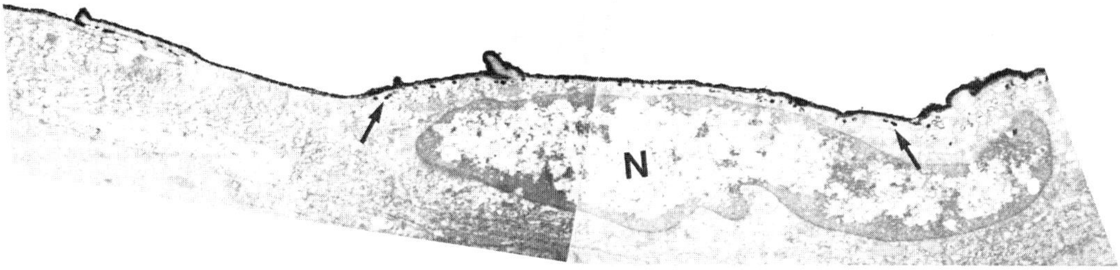

Fig. 23. Conjunctival epithelium of a cynomolgus monkey 5 min after application of HP in the conjunctival sac. The surface of the epithelium is coated with a thin, electron-dense layer of reaction product. Note the many pinocytotic vesicles (arrows). N = Cell nucleus. × 11,400.

still concentrated in the surface cytoplasm. Vesicles lying deeper in the epithelial cytoplasm were usually larger than those on the surface, meaning that numerous vesicles had coalesced during migration through the cell. Noteworthy was a diffuse storage of HP by some cells after 30 min. These were for the most part mitochondria-rich type V cells. The surface cells contained a large number of small and large pinocytotic vesicles after 60 min of incubation; these were filled with reaction product. Additionally, however, a certain amount of reaction product was found within the intercellular spaces, although the junctional complexes remained unchanged.

4.1.2.3. HP Resorption in Rabbit Conjunctiva

The process of HP resorption in rabbit conjunctiva was essentially the same as in monkey conjunctiva, although it was more rapid. After 5 min, the type V cells of the rabbit conjunctiva had already stored HP diffusely. Large vesicles had already formed in the other cells by this time and had penetrated to the deep layers of cytoplasm in the surface epithelial cell layer. After 30 min, reaction product was present in the apical sections of the intercellular spaces distally from the junctional complexes, that is, the pinocytotic vesicles had already migrated throughout the entire cytoplasm of the surface epithelial cells (fig. 24). After 60 min, the basal portions of the intercellular spaces of rabbit conjunctiva were filled with reaction product. At this point HP was not present in either monkey or rabbit subepithelial conjunctival tissue or in the lumina of the capillaries.

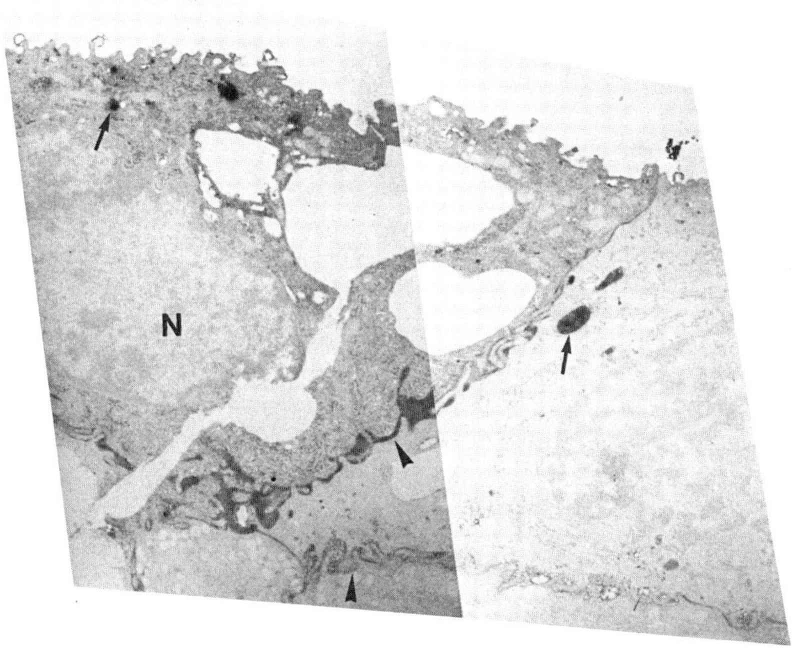

Fig. 24. Conjunctival epithelium of an albino rabbit 30 min after HP was placed in the conjunctival sac. Reaction product is visible both in the vesicles within the cytoplasm (arrows) and in the intercellular spaces (arrowheads). N = Cell nucleus. × 10,600.

4.1.3. Effect of an Antiglaucomatous Agent (Pilocarpine) on the Conjunctival Ultrastructure of Cynomolgus Monkeys

The experiment produced an almost immediate increase in secretory activity in the lid apparatus of the 2 monkeys given pilocarpine for 3 months from a medicine dispenser. Evaluation of the epithelial surface cells of the conjunctiva showed that essentially the same cell types were present in both animals as in normal monkey conjunctiva. However, the respective percentages of the individual cell types had changed considerably (fig. 25 a–c). Electron microscopic analysis of the tissue specimens after 3 months showed that nearly all goblet cells had vanished. They were replaced by type II cells, whose numbers increased considerably. Likewise the numbers of electron-dense granules present in the surface cytoplasm of these cells were markedly increased (fig. 26). In addition to the granules predominat-

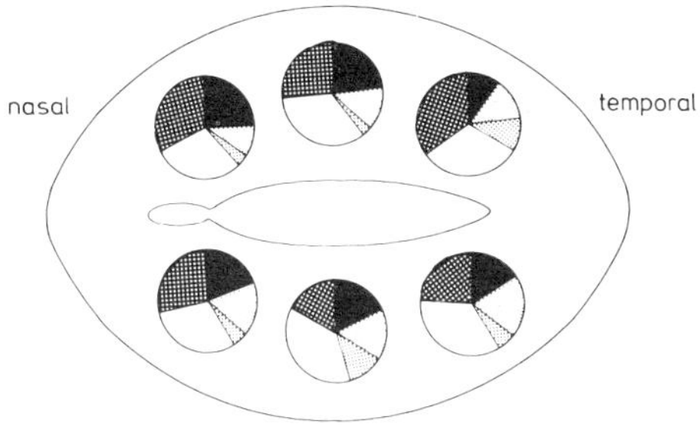

Conj. tarsi	Type I	Type II	Type III	Type IV	Type V
Nasal lower lid	19	17	5	30	29
Temporal lower lid	16	19	6	35	24
Central lower lid	18	16	11	38	17
Nasal upper lid	24	10	4	29	33
Temporal upper lid	10	13	10	32	35
Central upper lid	23	13	3	34	27

a

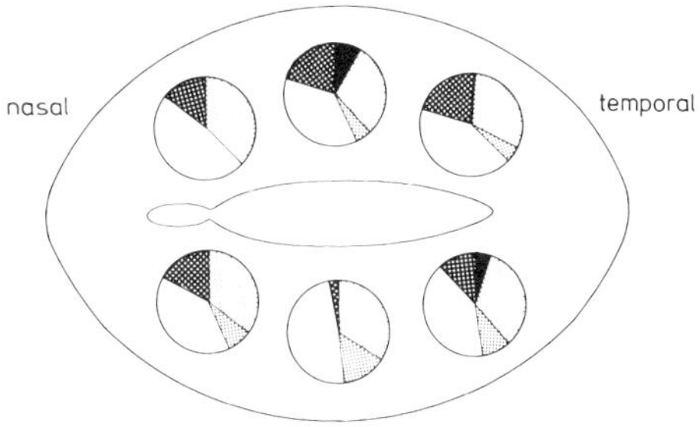

Conj. tarsi	Type I	Type II	Type III	Type IV	Type V
Nasal lower lid	0	35	8	40	17
Temporal lower lid	5	33	9	41	12
Central lower lid	0	34	14	49	3
Nasal upper lid	0	37	0	49	14
Temporal upper lid	1	31	5	43	20
Central upper lid	8	30	5	37	20

b

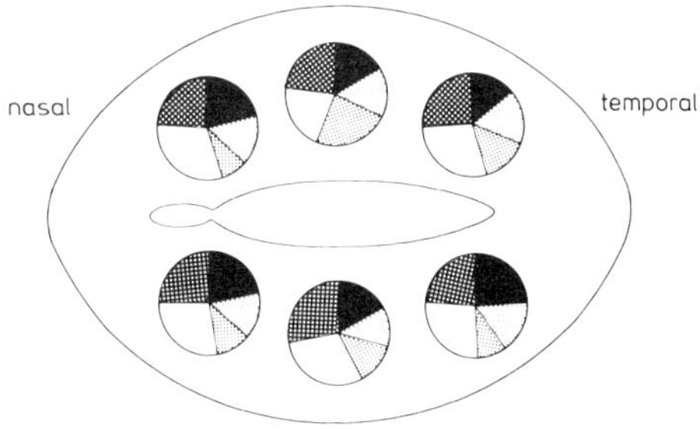

Conj. tarsi	Type I	Type II	Type III	Type IV	Type V
Nasal lower lid	22	14	11	28	25
Temporal lower lid	24	16	9	27	24
Central lower lid	17	12	13	30	28
Nasal upper lid	21	16	8	31	24
Temporal upper lid	14	17	14	29	26
Central upper lid	17	15	24	21	23

c

Fig. 25. Distribution of the 5 different cell types by percent among epithelial surface cells in tarsal conjunctiva of the cynomolgus monkey. *a* Control. *b* After pilocarpine application by a medicine-dispensing implant. *c* After application of 4% pilocarpine eye drops.

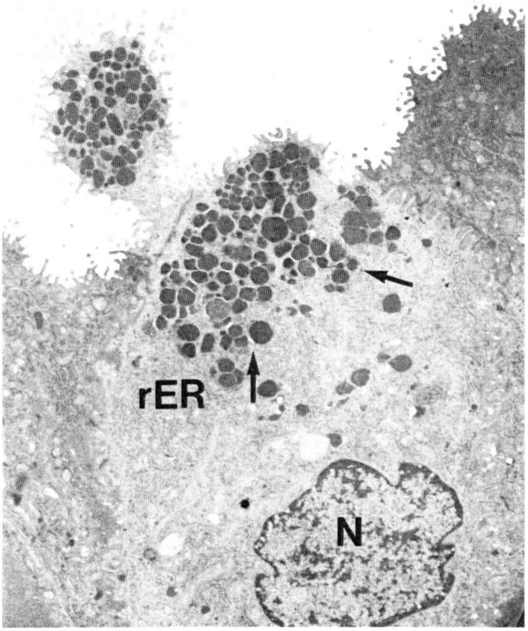

Fig. 26. Electron micrograph of a type II cell after pilocarpine application by a medicine dispenser. Note the large number of secretory granules in the cell's apical cytoplasm (arrows). rER = Rough endoplasmatic reticulum; N = cell nucleus. × 5,100.

ing in the cytoplasm, the cells contained well-developed rER and numerous mitochondria (fig. 27).

The application of pilocarpine drops had almost no effect on the distribution of the 5 cell types in the conjunctiva in comparison to the controls (fig. 25). However, the epithelial cells had an increased number of Golgi complexes following local administration of pilocarpine.

4.1.4. Corneal Reepithelialization at the Rabbit Limbus: Ultrastructure and Histochemistry

The epithelial surface cells of the limbus region were flat near the cornea and cubical near the area of transition to bulbar conjunctiva (fig. 28 a, b). The flat and cubical cells contained an unusually large number of mitochondria, so that they were for the most part to be termed type V cells.

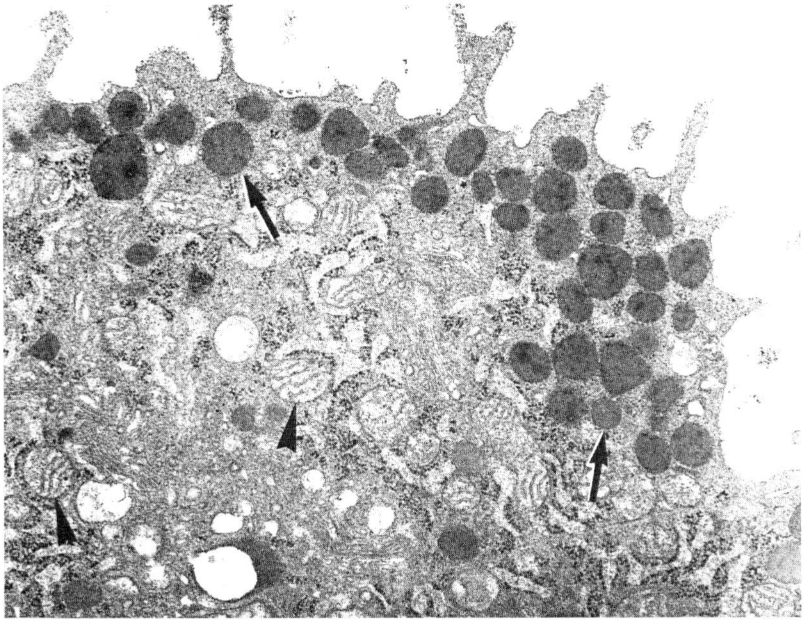

Fig. 27. Electron micrograph of a type II cell after pilocarpine application by a medicine dispenser. Arrows = Secretory granules; arrowheads = mitochondria. × 19,200.

Moreover, other small vesicles were present in the apical cytoplasm, along with sections of rER and Golgi complexes. In some cells, the Golgi complex was so dominant that they had to be classified as type III cells. There were only scattered ergastoplasm-rich (type IV) cells. We found no goblet (type I) cells or cells with osmiophilic granules (type II).

No significant differences were found among the 5 cell types themselves or in their distribution in the nasal, temporal, superior and inferior sections of the limbus epithelium [111]. Noteworthy, however, was the fact that the rabbit limbal epithelium – in contrast to the tarsal conjunctiva – contained numerous tonofilaments; these, with the desmosomes, strengthened intercellular contact. This tonofilament content increased considerably toward the corneal epithelium. The histochemical carbonic anhydrase reaction test showed that this enzyme was present only in the basal cells of the limbus epithelium. The cytoplasm, the cell membrane and the cell

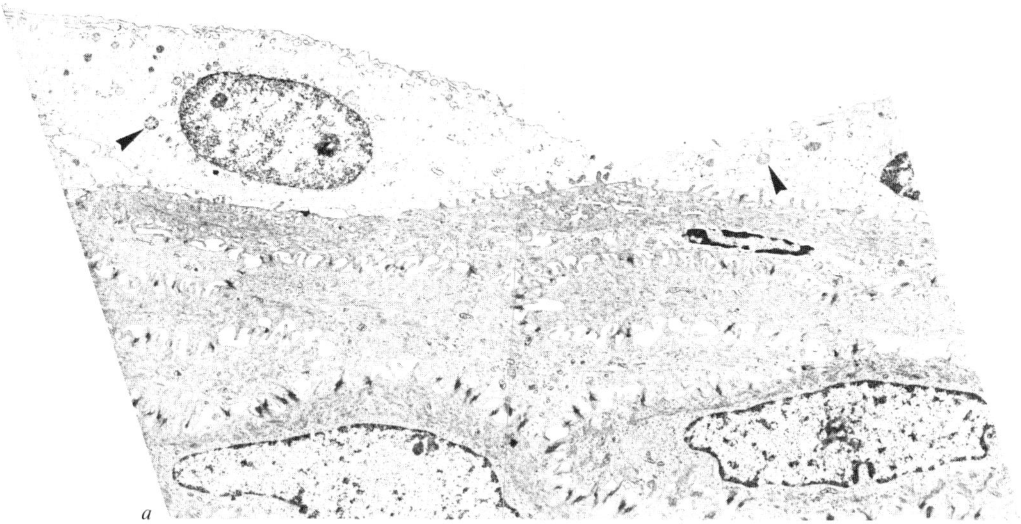

a

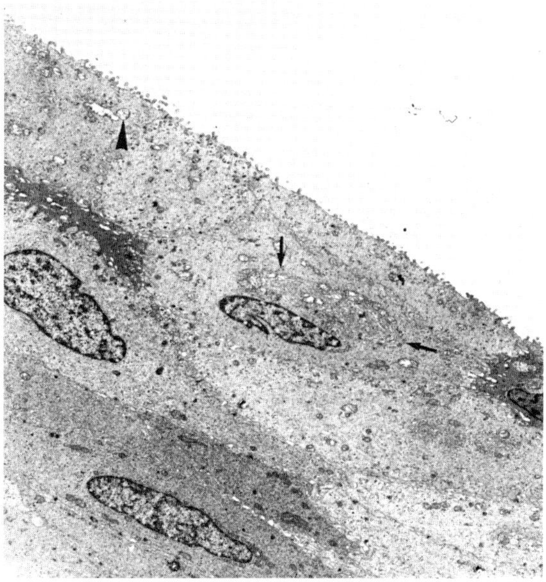

b

Fig. 28. a, b Electron micrograph of limbus epithelium of an albino rabbit near the cornea *(a)* and remote from it *(b)*. Arrows = Golgi complex; arrowheads = mitochondria. $a \times 4{,}320$; $b \times 3{,}000$.

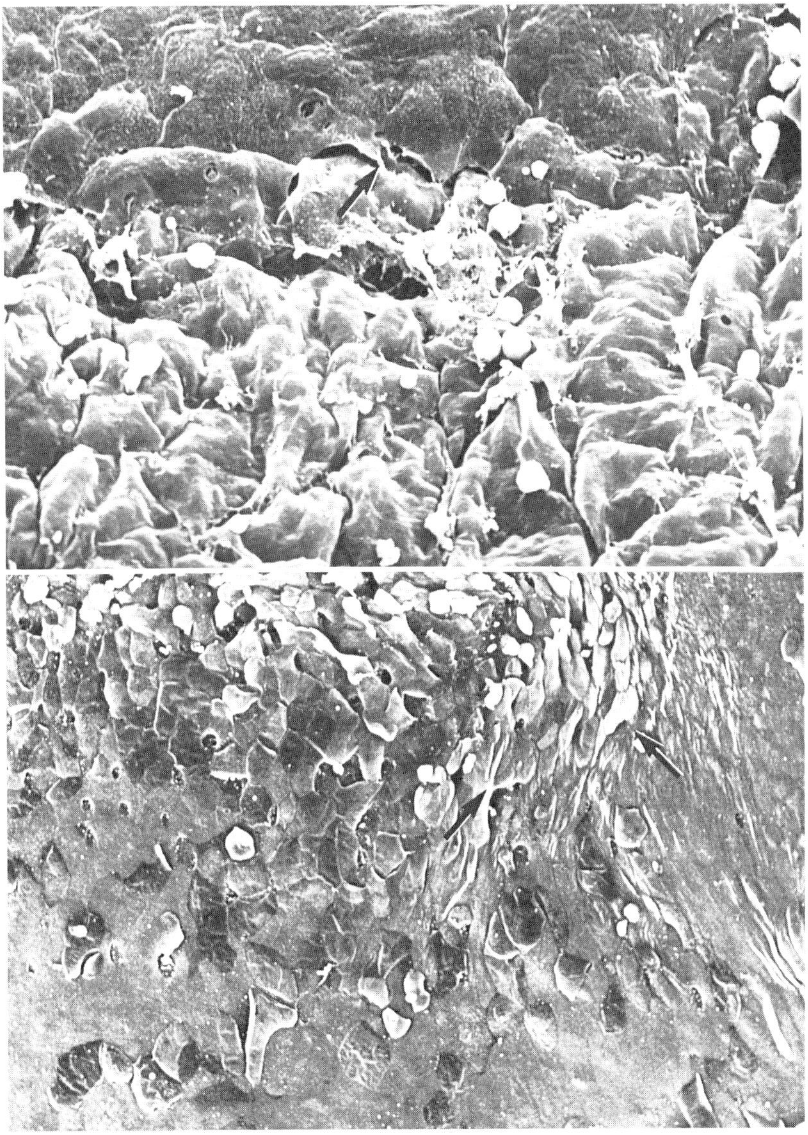

a

b

Fig. 29. a, b SEM view of the corneal surface 12 h *(a)* and 24 h *(b)* after abrasion. *a* Note the cytoplasmic extensions of the epithelial cells proliferating forward onto the basement membrane (arrow). *b* After 24 h a strand of epithelial cells has proliferated out onto the cornea (arrows). *a* × 1,120; *b* × 452.

nucleus of the basal limbal epithelial cells reacted positively to the carbonic anhydrase test [111]. Neither the adjacent corneal epithelium nor the bulbar conjunctiva was stainable.

Twelve hours after reepithelialization began, about 0.5 mm of the abraded cornea were reepithelialized from the limbus outwards. The cells with farthest forward proliferation at this point showed pseudopod-like cytoplasmatic extensions on the corneal basement membrane. There were no microvilli (fig. 29a). In the TEM micrograph, the forward-proliferating epithelial cells showed a relatively low tonofilament content in contrast to the corneal epithelium, whereas many mitochondria were present. All forward proliferating epithelial cells stained positively with the carbonic anhydrase reaction test (fig. 30a).

After 24 h the SEM micrograph showed a strand-like forward proliferation of epithelial surface cells onto the cornea (fig. 29 b). About 1.7 mm of the corneal periphery were now covered by a single layer of cells, and a few individual cells had proliferated even farther forward. In contrast to normal corneal epithelial cells, the epithelial cells with farthest forward proliferation still contained relatively few tonofilaments and, in contrast to normal limbal epithelial cells, a reduced number of mitochondria. In the basal cell area there were circumscribed concentrations in the basement membrane and tonofilament clusters which corresponded to emergent hemidesmosomes (fig. 31) [181]. The cell membrane, the cell nucleus and the cytoplasm of the forward-proliferating epithelial cells all stained histochemically positive. By this time the basal epithelial cells in the limbus area stained clearly less than those of the control animal.

After 48 h only small areas of the central cornea remained nonepithelialized. In the TEM micrograph, the forward-proliferating epithelial cells already had numerous desmosomes and an abundance of tonofilaments. There were typical hemidesmosomes near the basement membrane, and tight junctions now appeared for the first time towards the surface of the epithelium (fig. 32 a, b).

Histochemically, only scattered epithelial cells which had proliferated forward onto the cornea near the limbus stained positive (fig. 30 b). In contrast, the limbus region showed the typical, pronounced reaction to carbonic anhydrase staining (fig. 30 c).

After 7 days the surface of the cornea was reepithelialized and corresponded morphologically and histochemically to that of the control animal. Only one laboratory animal still had a reduced content of microvilli in the SEM photo.

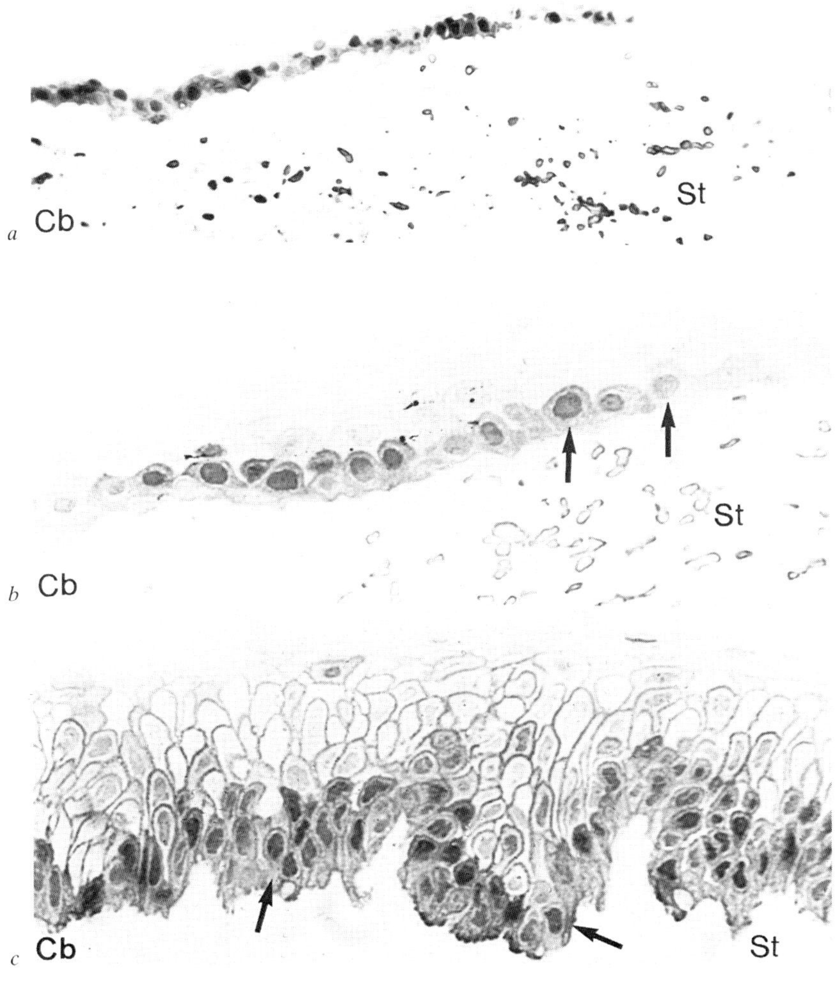

Fig. 30. a–c Light microscopic view of regenerating corneal epithelium after carbonic anhydrase staining. Cb = Transition of the bulbar conjunctiva; St = corneal stroma. *a* 12 h after abrasion. All forward-proliferating cells stain positive. × 190. *b* 48 h after abrasion. Only few epithelial cells with forward proliferation onto the cornea stain positive (arrows). × 244. *c* Carbonic anhydrase test in the limbal epithelium 48 h after abrasion. Note the especially intense staining of the basal epithelial cells (arrows). × 244.

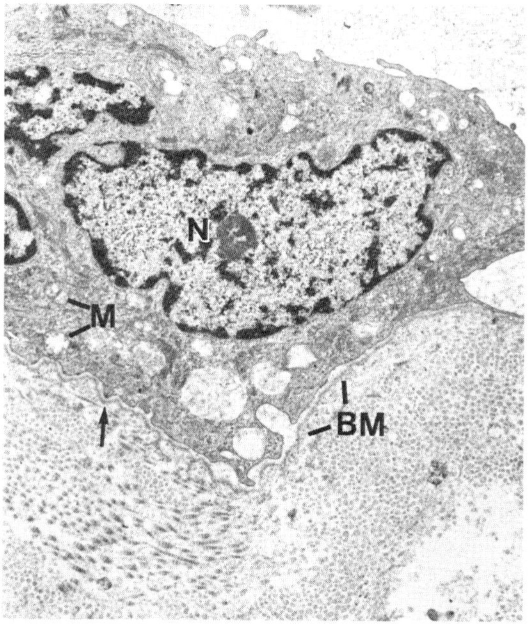

Fig. 31. Electron micrograph of the epithelial cell with farthest forward proliferation 24 h after abrasion. N = Cell nucleus; M = mitochondria; BM = basement membrane; arrow = emerging hemidesmosomes. × 6,400.

4.2. Human Conjunctiva

4.2.1. Ultrastructure of Healthy Conjunctival Epithelium: Age-Related and Regional Differences

Epithelial surface cells of human conjunctiva are cubical on the tarsus (fig. 33) but highly prismatic in the fornix area. The epithelium of bulbar conjunctiva, on the other hand, is flatter, resembling tarsal conjunctiva in this respect. Bulbar conjunctiva then merges into the limbal epithelium, where epithelial surface cells are visibly flatter and finally accommodate themselves seamlessly to the corneal epithelium.

As with the conjunctiva of the laboratory animals studied, the intercellular spaces were completely closed in the direction of the surface by tight junctions (fig. 34). Extensions of cytoplasm extended into the intercellular spaces, sometimes connected by occasional desmosomes. These microvilli were also present on the cell surface of the epithelial cells.

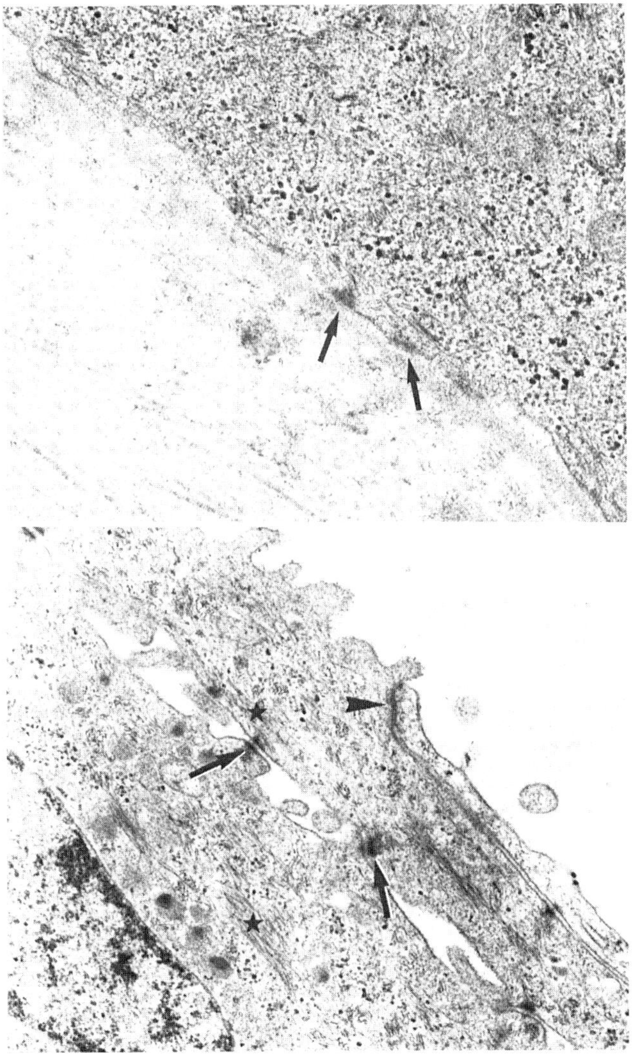

Fig. 32. a, b Electron micrograph of the forward-proliferating limbus epithelium in the middle corneal periphery 48 h after abrasion. *a* Basal epithelium. Note the hemidesmosomes, now clearly recognizable (arrows). *b* Apical part of the epithelium with numerous tonofilaments (stars), desmosomes (arrows) and a tight junction (arrowhead). *a* × 24,500; *b* × 27,200.

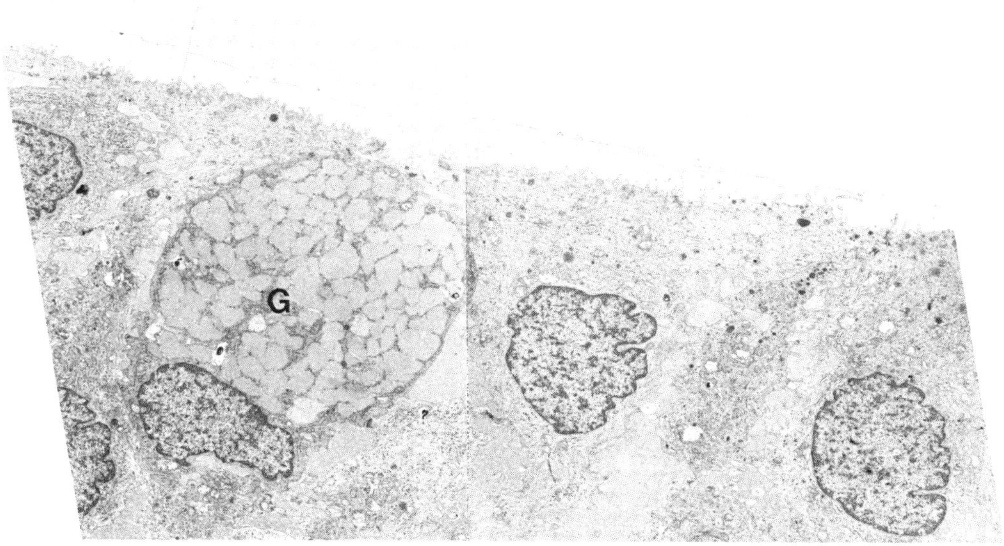

Fig. 33. Electron microscopic overview of tarsal conjunctiva. Note the cubical epithelial surface cells. G = Goblet cells. Patient K.B., 25 years old, temporal upper lid. × 3,350.

Several microvilli often coalesced to form large microplicae. The basal epithelial cells clung with hemidesmosomes to a quite thick basement membrane. Small, fenestrated capillaries extended also in humans almost to this basement membrane.

The epithelial surface cells of human conjunctiva can be assigned to different cell types according to the criteria described earlier (4.1.1.1.), although the type II cells in human conjunctiva were composed not only of cells with osmiophilic granules (fig. 35a) but also of cells with ca. 100–500 nm large vesicles. The content of these vesicles appeared inhomogeneously osmiophilic. In the apical area of the cells, the vesicles partly coalesced with the cell membrane and released either their content alone or their surrounding membrane as well to the surface of the epithelium (fig. 35b, c). The vesicles' origin in Golgi apparatus was clearly recognizable in many epithelial cells. These cells thus exhibited an unmistakable merocrine type of secretion, whereas an apocrine secretion mechanism was found among type II cells in the fornix conjunctivae with highly prismatic cells. Here there were individual cells which squeezed out whole units of cytoplasm, thus releasing intracellularly formed products to their surroundings (fig. 35 d).

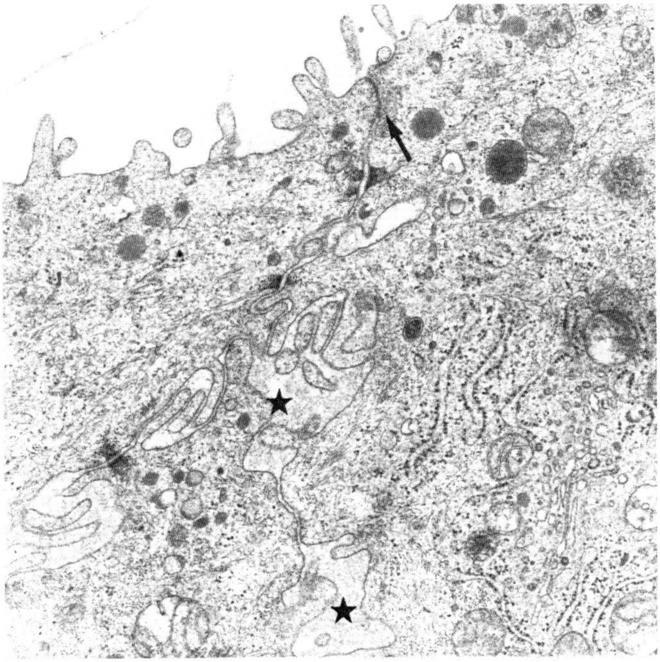

Fig. 34. The apical area of 2 surface epithelial cells of the conjunctival fornix. The intercellular space (stars) is closed toward the surface of the epithelium by tight junctions (arrow). Patient M.B., 63 years, nasal upper lid. × 17,600.

On the whole, the type II cells with their inhomogeneous, osmiophilic, somewhat granular contents were the most common. They were found in varying numbers in all regions studied in the human conjunctiva. Second most common were the cells with homogeneous osmiophilic granules, which also appeared as type II cells more or less often in all regions of the epithelium. The apocrine mode of secretion was found only in conjunctival areas with multilayered cylinder epithelium, that is, in the fornix, where the other two types of type II cells described above were also present.

Worthy of note was the varying structure of the cytoskeleton in epithelial surface cells of the different conjunctival regions. This intracytoplasmatic system of protein filaments was almost nonexistent in the conjunctival fornix. The tarsal and bulbar conjunctiva, on the other hand, had considerably more desmosomes, interconnected by intracellular

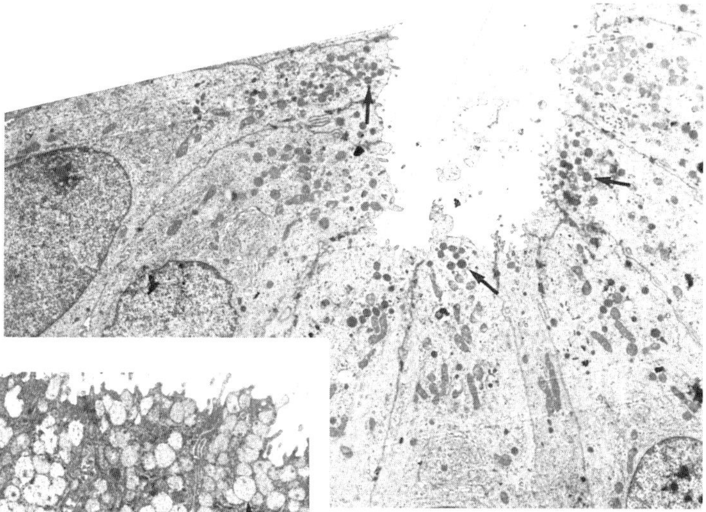

35a

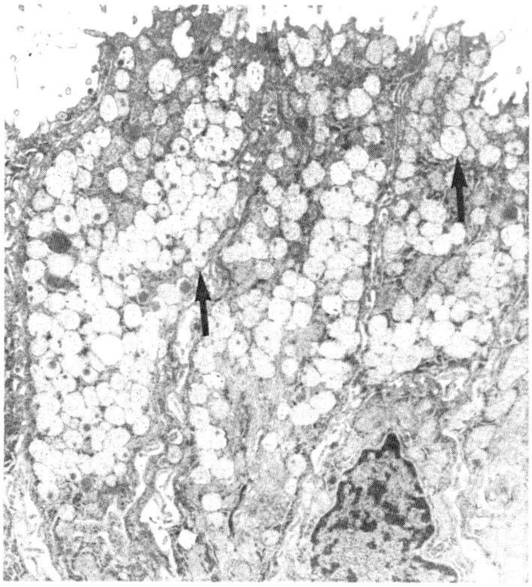

35b

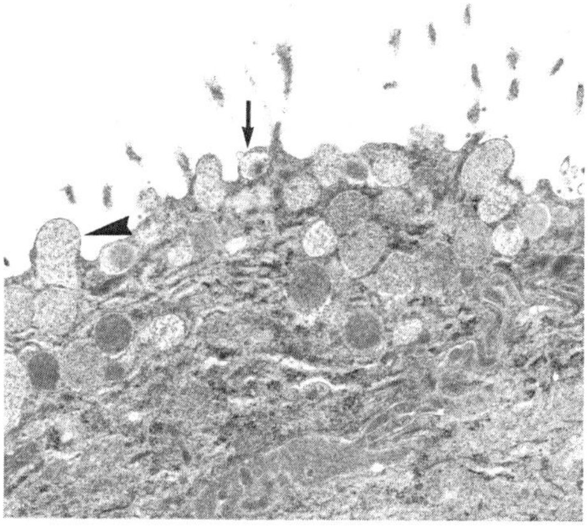

35c

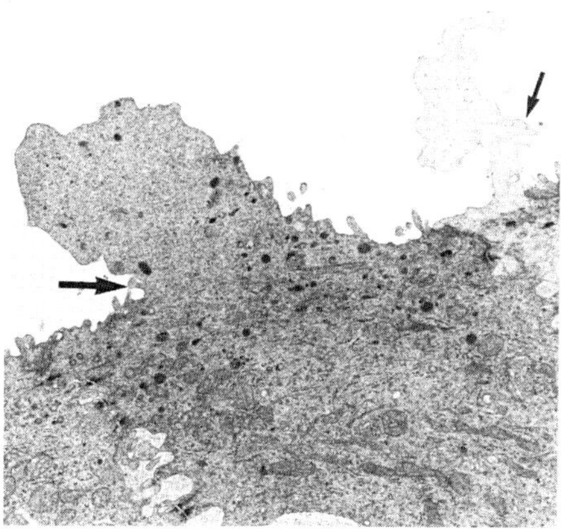

35d

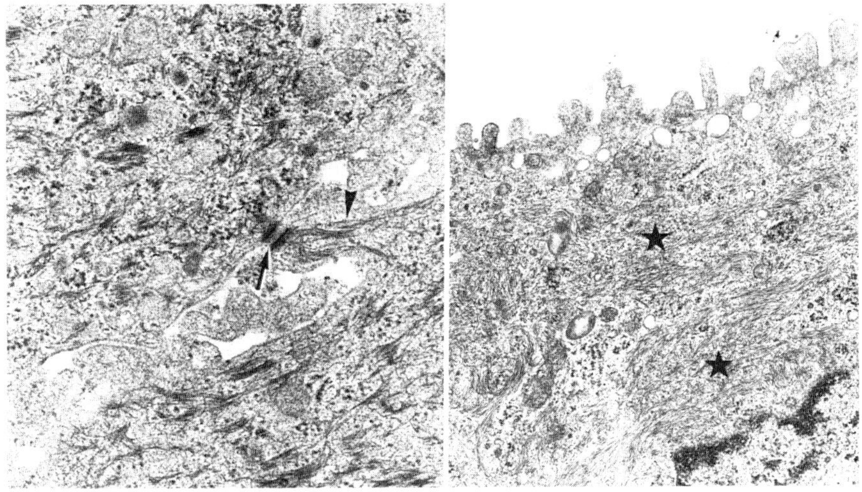

a b

Fig. 36. a, b Electron micrograph of two epithelial cells of bulbar conjunctiva. *a* Thick bundles of tonofilaments (arrowhead) extend from the desmosomes (arrow), which mechanically bind the plasma membranes of neighboring cells, into the cytoplasm. Patient N. Sch., 4 years, nasal inferior region. × 26,680. *b* Apical cell region with numerous tonofilaments (stars). Patient G. G., 46 years, nasal superior region. × 17,000.

tonofilaments (fig. 36a, b). In the limbus area, nearly the entire cytoplasm of the epithelial surface cells was filled by elements of the cytoskeleton, so that the organelles here and thus the distinction between cells played a subordinate role.

The distribution of the 5 different cell types remained relatively constant in the persons studied and showed only slight dependency on age.

Fig. 35. a–d Electron micrographs of type II cells of human conjunctiva. *a* Highly prismatic type II cells with homogeneous, osmiophilic granules (arrows). Patient E. W., 23 years, fornix conjunctivae, temporal lower lid. × 3,330. *b* Type II cells with inhomogeneous osmiophilic granules. The secretory granules are thickly clustered in the apical cell area (arrows). Patient J. L., 52 years, tarsal conjunctiva, central upper lid. × 6,500. *c* Individual vesicles are incorporated into the cell membrane (arrow); others are passed on to the surface of the epithelium (arrowhead). Patient I. H., 78 years, tarsal conjunctiva, nasal upper lid. × 12,500. *d* Apocrine mechanism of secretion. Whole units of cytoplasm are squeezed off on the cell surface (arrows). Patient M. B., 63 years. Fornix conjunctivae, temporal upper lid. × 5,150.

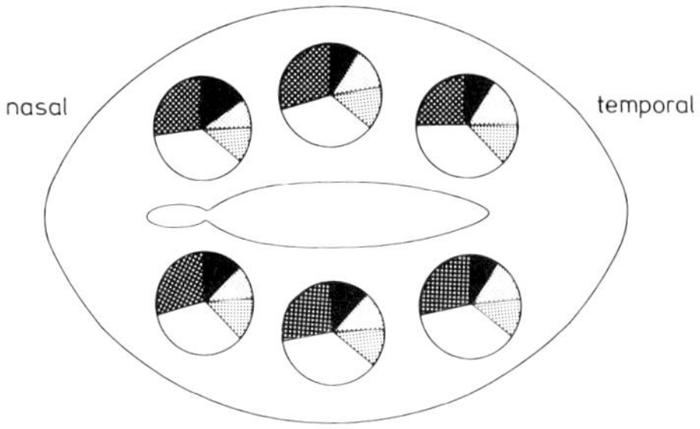

Conj tarsi	Type I	Type II	Type III	Type IV	Type V
Nasal lower lid	$\bar{X}$: 12.75 S: 3.59	$\bar{X}$: 10.75 S: 5.91	$\bar{X}$: 13.5 S: 1.29	$\bar{X}$: 34 S: 3.56	$\bar{X}$: 29 S: 4.24
Temporal lower lid	$\bar{X}$: 9.25 S: 5.26	$\bar{X}$: 14.5 S: 2.89	$\bar{X}$: 11.25 S: 0.96	$\bar{X}$: 36.25 S: 6.45	$\bar{X}$: 28.75 S: 5.5
Central lower lid	$\bar{X}$: 11.75 S: 4.79	$\bar{X}$: 12.75 S: 4.50	$\bar{X}$: 11.5 S: 1.73	$\bar{X}$: 36.5 S: 6.46	$\bar{X}$: 27.5 S: 3.87
Nasal upper lid	$\bar{X}$: 14.5 S: 1.29	$\bar{X}$: 10.5 S: 6.19	$\bar{X}$: 11.5 S: 3.51	$\bar{X}$: 36.75 S: 3.78	$\bar{X}$: 26.75 S: 1.71
Temporal upper lid	$\bar{X}$: 9 S: 4.97	$\bar{X}$: 15.5 S: 3.42	$\bar{X}$: 12.25 S: 2.22	$\bar{X}$: 38 S: 4.55	$\bar{X}$: 25.25 S: 4.19
Central upper lid	$\bar{X}$: 9.75 S: 3.30	$\bar{X}$: 13 S: 4.08	$\bar{X}$: 13.25 S: 2.5	$\bar{X}$: 35.75 S: 6.24	$\bar{X}$: 28.25 S: 2.22

a

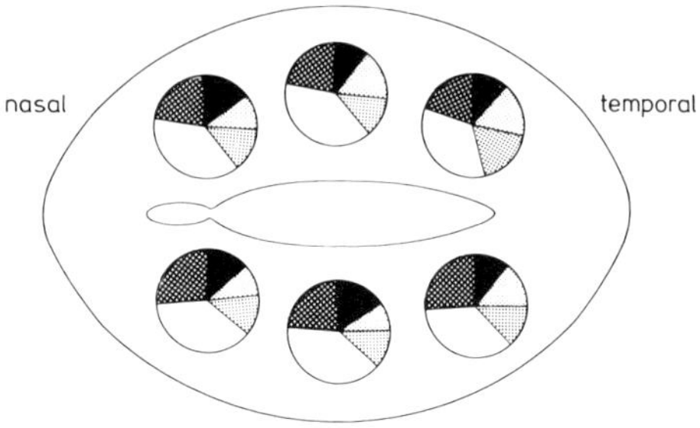

Conj. tarsi	Type I	Type II	Type III	Type IV	Type V
Nasal lower lid	$\bar{X}$: 14 S: 2.31	$\bar{X}$: 9.5 S: 1.73	$\bar{X}$: 12.5 S: 3.87	$\bar{X}$: 37.75 S: 4.57	$\bar{X}$: 26.25 S: 1.5
Temporal lower lid	$\bar{X}$: 10.25 S: 4.72	$\bar{X}$: 14.75 S: 3.30	$\bar{X}$: 13.25 S: 1.92	$\bar{X}$: 35.75 S: 3.20	$\bar{X}$: 26 S: 3.46
Central lower lid	$\bar{X}$: 16 S: 6.06	$\bar{X}$: 8.25 S: 3.86	$\bar{X}$: 12.75 S: 5.32	$\bar{X}$: 39 S: 4.32	$\bar{X}$: 24 S: 1.41
Nasal upper lid	$\bar{X}$: 14.25 S: 3.86	$\bar{X}$: 11.5 S: 5.32	$\bar{X}$: 13.25 S: 2.87	$\bar{X}$: 38.5 S: 4.80	$\bar{X}$: 22.5 S: 3.79
Temporal upper lid	$\bar{X}$: 11.75 S: 2.99	$\bar{X}$: 16.5 S: 2.65	$\bar{X}$: 17 S: 2.16	$\bar{X}$: 35.25 S: 6.5	$\bar{X}$: 19.5 S: 1.73
Central upper lid	$\bar{X}$: 10.25 S: 2.36	$\bar{X}$: 16.25 S: 2.22	$\bar{X}$: 12.25 S: 2.75	$\bar{X}$: 39.25 S: 3.20	$\bar{X}$: 22 S: 4.83

b

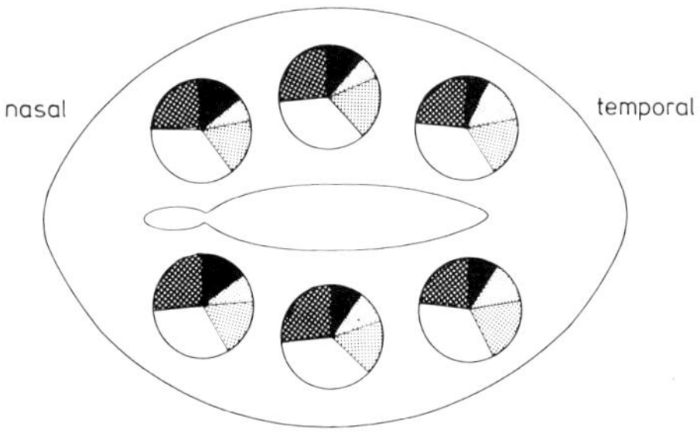

Conj tarsi	Type I	Type II	Type III	Type IV	Type V
Nasal lower lid	$\bar{X}$: 15 S: 2.58	$\bar{X}$: 8.75 S: 2.50	$\bar{X}$: 17.75 S: 2.38	$\bar{X}$: 31.5 S: 3.30	$\bar{X}$: 27 S: 5.48
Temporal lower lid	$\bar{X}$: 9 S: 2.94	$\bar{X}$: 13.25 S: 5.74	$\bar{X}$: 20.25 S: 2.99	$\bar{X}$: 34.25 S: 2.75	$\bar{X}$: 23.25 S: 3.86
Central lower lid	$\bar{X}$: 9.5 S: 3.87	$\bar{X}$: 9.75 S: 4.35	$\bar{X}$: 17.25 S: 3.95	$\bar{X}$: 36.5 S: 4.93	$\bar{X}$: 27 S: 3.37
Nasal upper lid	$\bar{X}$: 14 S: 1.63	$\bar{X}$: 7.5 S: 3.70	$\bar{X}$: 17.5 S: 2.65	$\bar{X}$: 37 S: 4.69	$\bar{X}$: 24 S: 2.94
Temporal upper lid	$\bar{X}$: 6.75 S: 4.43	$\bar{X}$: 14.75 S: 9.36	$\bar{X}$: 19.25 S: 5.19	$\bar{X}$: 36.75 S: 3.78	$\bar{X}$: 22.5 S: 6.70
Central upper lid	$\bar{X}$: 12 S: 5.35	$\bar{X}$: 7.25 S: 3.30	$\bar{X}$: 19.25 S: 4.76	$\bar{X}$: 35.75 S: 4.5	$\bar{X}$: 25.75 S: 4.5

c

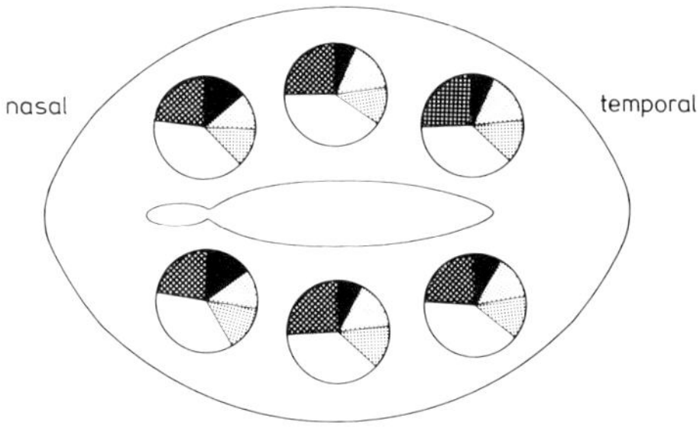

Conj. tarsi	Type I	Type II	Type III	Type IV	Type V
Nasal lower lid	$\bar{X}$: 15 S: 4.32	$\bar{X}$: 12.25 S: 2.5	$\bar{X}$: 14.75 S: 2.87	$\bar{X}$: 35.75 S: 4.27	$\bar{X}$: 22.25 S: 1.71
Temporal lower lid	$\bar{X}$: 8.25 S: 2.06	$\bar{X}$: 14.25 S: 2.87	$\bar{X}$: 13 S: 2.94	$\bar{X}$: 40.75 S: 5.06	$\bar{X}$: 23.75 S: 4.57
Central lower lid	$\bar{X}$: 7.5 S: 1.92	$\bar{X}$: 15.5 S: 4.66	$\bar{X}$: 13.25 S: 4.86	$\bar{X}$: 37.5 S: 3.87	$\bar{X}$: 26.25 S: 3.30
Nasal upper lid	$\bar{X}$: 14 S: 3.56	$\bar{X}$: 11.75 S: 2.63	$\bar{X}$: 12.25 S: 2.06	$\bar{X}$: 39.5 S: 5.07	$\bar{X}$: 22.5 S: 4.04
Temporal upper lid	$\bar{X}$: 7.5 S: 3.0	$\bar{X}$: 16.25 S: 2.63	$\bar{X}$: 13.25 S: 1.26	$\bar{X}$: 37.75 S: 2.89	$\bar{X}$: 25.25 S: 1.71
Central upper lid	$\bar{X}$: 6.75 S: 2.87	$\bar{X}$: 16 S: 3.56	$\bar{X}$: 11.75 S: 2.75	$\bar{X}$: 40.75 S: 4.35	$\bar{X}$: 24.75 S: 5.38

d

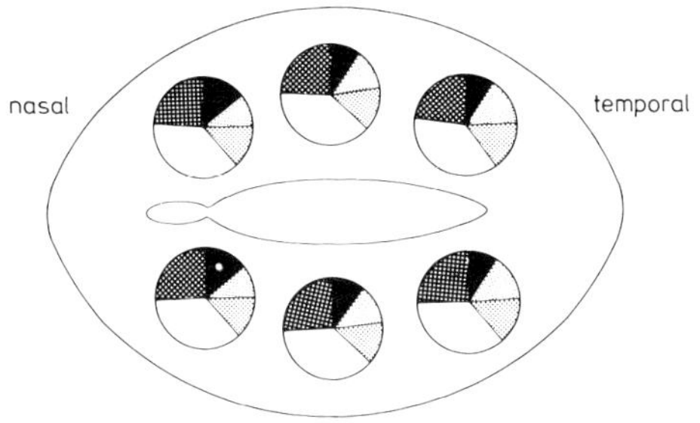

e

Conj. tarsi	Type I	Type II	Type III	Type IV	Type V
Nasal lower lid	X̄: 14.19 S: 3.10	X̄: 10.31 S: 3.46	X̄: 14.56 S: 3.14	X̄: 34.69 S: 4.32	X̄: 26.13 S: 4.13
Temporal lower lid	X̄: 9.25 S: 3.62	X̄: 14.19 S: 3.53	X̄: 14.50 S: 4.12	X̄: 36.75 S: 4.82	X̄: 25.44 S: 4.55
Central lower lid	X̄: 11.19 S: 5.13	X̄: 11.56 S: 4.84	X̄: 13.69 S: 4.36	X̄: 37.38 S: 4.59	X̄: 26.19 S: 3.13
Nasal upper lid	X̄: 14.19 S: 2.54	X̄: 10.31 S: 4.53	X̄: 13.63 S: 3.48	X̄: 37.94 S: 4.28	X̄: 23.94 S: 3.42
Temporal upper lid	X̄: 8.75 S: 4.04	X̄: 15.75 S: 4.81	X̄: 15.44 S: 4.02	X̄: 36.89 S: 4.27	X̄: 23.19 S: 4.43
Central upper lid	X̄: 9.69 S: 3.81	X̄: 13.13 S: 4.80	X̄: 14.06 S: 4.22	X̄: 37.88 S: 4.77	X̄: 25.19 S: 4.56

Fig. 37. Distribution of the 5 cell types in the regions studied in human tarsal conjunctiva in percent at various patient ages. X̄ = Mean; S = standard deviation. *a* 1–20 years. *b* 21–40 years. *c* 41–60 years. *d* > 60 years. *e* Independently of age.

4.2.1.1. Tarsal Conjunctiva (fig. 37a–e)

Type IV (ergastoplasm-rich) cells were most frequent in the tarsal conjunctiva; they constituted 35–40% of epithelial surface cells, slightly more in subjects over 60 years old (fig. 40).

Epithelial cells with numerous mitochondria (type V) were the second most common. They constituted 25–30% of cells in the conjunctival epithelium of subjects 20 years old or younger, but 25% or less in older patients.

Type III cells (with a sharply defined Golgi complex) were slightly more frequent (just under 20%) among patients between 40 and 60 years of age, but comprised only 15% of epithelial surface cells among those

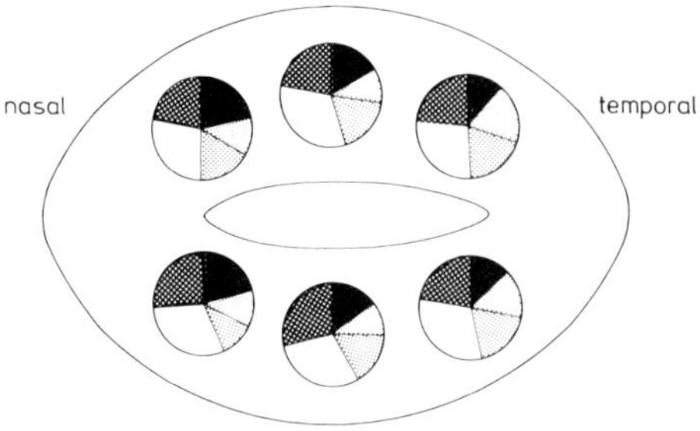

Fornix	Type I	Type II	Type III	Type IV	Type V
Nasal lower lid	X̄: 21.25 S: 4.99	X̄: 11.25 S: 6.19	X̄: 11 S: 3.16	X̄: 30.5 S: 2.52	X̄: 26 S: 6.06
Temporal lower lid	X̄: 13.25 S: 2.63	X̄: 14.75 S: 3.5	X̄: 17.75 S: 3.4	X̄: 31.5 S: 7.14	X̄: 22.75 S: 6.29
Central lower lid	X̄: 15.25 S: 2.06	X̄: 10.75 S: 5.19	X̄: 16.25 S: 7.23	X̄: 29.5 S: 3.70	X̄: 28.25 S: 7.41
Nasal upper lid	X̄: 21.5 S: 2.08	X̄: 12 S: 6.58	X̄: 15.75 S: 4.99	X̄: 28.5 S: 8.10	X̄: 22.25 S: 6.24
Temporal upper lid	X̄: 11.75 S: 3.95	X̄: 18 S: 2.94	X̄: 18.75 S: 5.32	X̄: 27.75 S: 10.63	X̄: 23.75 S: 4.99
Central upper lid	X̄: 16.75 S: 2.22	X̄: 11.25 S: 7.89	X̄: 17.25 S: 2.87	X̄: 32.5 S: 7.05	X̄: 22.25 S: 7.50

a

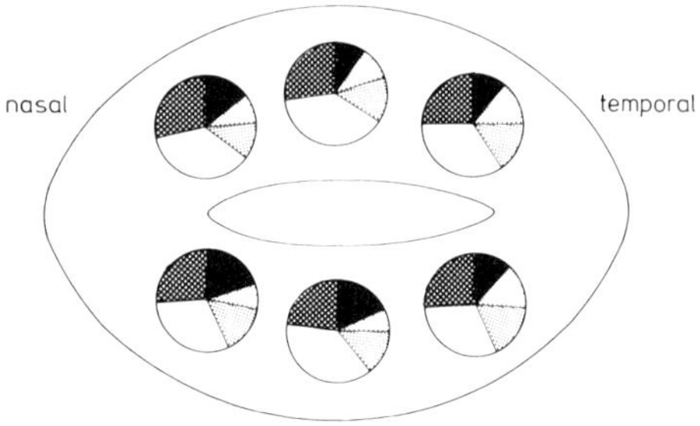

Fornix	Type I	Type II	Type III	Type IV	Type V
Nasal lower lid	X̄: 19.25 S: 0.96	X̄: 8.75 S: 3.78	X̄: 15 S: 4.32	X̄: 31.75 S: 1.26	X̄: 25.25 S: 5.56
Temporal lower lid	X̄: 12.25 S: 4.19	X̄: 13.75 S: 2.75	X̄: 16.75 S: 2.87	X̄: 31 S: 4	X̄: 26.25 S: 4.03
Central lower lid	X̄: 17.75 S: 3.40	X̄: 7.5 S: 3.11	X̄: 13.5 S: 3.70	X̄: 38 S: 2.58	X̄: 23.25 S: 2.06
Nasal upper lid	X̄: 14.5 S: 4.66	X̄: 9.5 S: 4.80	X̄: 11.25 S: 2.87	X̄: 36.25 S: 7.27	X̄: 28.5 S: 4.12
Temporal upper lid	X̄: 10.75 S: 2.87	X̄: 14 S: 4.69	X̄: 15.5 S: 2.52	X̄: 34.25 S: 1.71	X̄: 25.5 S: 2.38
Central upper lid	X̄: 9.5 S: 0.56	X̄: 10.75 S: 6.55	X̄: 13.75 S: 1.71	X̄: 38.75 S: 5.74	X̄: 27.25 S: 4.03

b

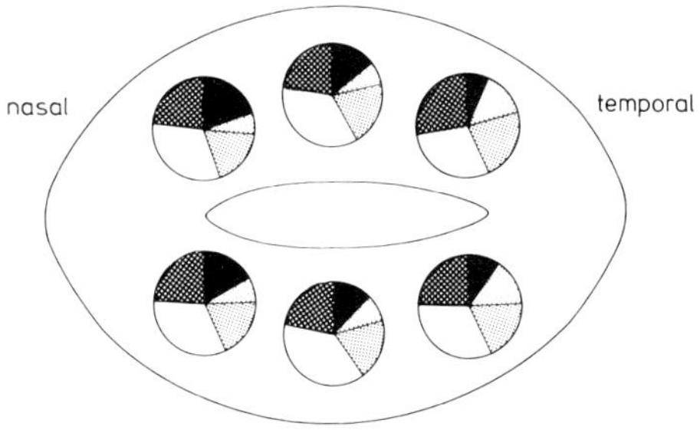

Fornix	Type I	Type II	Type III	Type IV	Type V
Nasal lower lid	X̄: 17 S: 1.41	X̄: 7.5 S: 3.11	X̄: 18.5 S: 3.87	X̄: 33 S: 5.16	X̄: 24 S: 3.16
Temporal lower lid	X̄: 9.5 S: 2.66	X̄: 14.25 S: 4.72	X̄: 19.25 S: 2.87	X̄: 31.75 S: 2.87	X̄: 25.25 S: 1.71
Central lower lid	X̄: 12 S: 3.83	X̄: 8.5 S: 3.11	X̄: 19 S: 1.83	X̄: 37.5 S: 5.0	X̄: 23 S: 3.16
Nasal upper lid	X̄: 19.75 S: 2.06	X̄: 8 S: 4.83	X̄: 16.75 S: 2.87	X̄: 31.75 S: 3.86	X̄: 23.75 S: 5.62
Temporal upper lid	X̄: 6.75 S: 0.96	X̄: 14.5 S: 4.57	X̄: 21.75 S: 0.96	X̄: 29.25 S: 6.89	X̄: 27.75 S: 6.95
Central upper lid	X̄: 14.25 S: 2.22	X̄: 7.5 S: 5.75	X̄: 20.25 S: 4.79	X̄: 35.25 S: 4.11	X̄: 22.75 S: 4.86

c

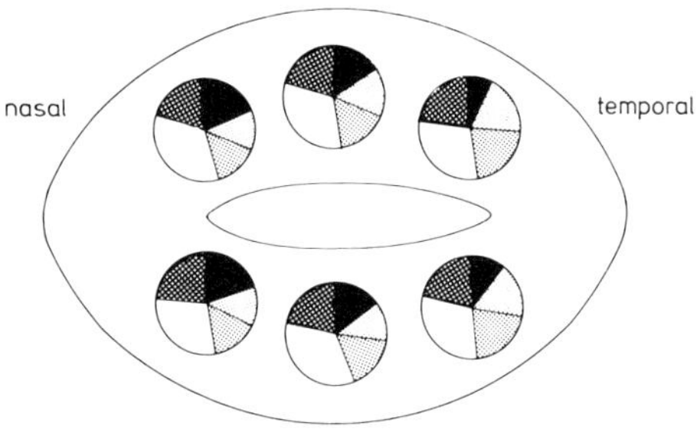

Fornix	Type I	Type II	Type III	Type IV	Type V
Nasal lower lid	X̄: 19.75 S: 3.30	X̄: 12.75 S: 4.79	X̄: 14.75 S: 2.36	X̄: 29.25 S: 2.06	X̄: 23.5 S: 5.45
Temporal lower lid	X̄: 10.5 S: 1.73	X̄: 16.75 S: 3.78	X̄: 21 S: 6.06	X̄: 30.25 S: 7.89	X̄: 21.5 S: 6.61
Central lower lid	X̄: 14.25 S: 3.40	X̄: 12.75 S: 4.79	X̄: 17 S: 3.92	X̄: 34.5 S: 6.35	X̄: 21.5 S: 3.87
Nasal upper lid	X̄: 18.75 S: 3.40	X̄: 12.5 S: 4.80	X̄: 14.5 S: 7.42	X̄: 34.0 S: 2.16	X̄: 20.25 S: 3.30
Temporal upper lid	X̄: 7.25 S: 2.50	X̄: 19.25 S: 6.29	X̄: 19.5 S: 5.32	X̄: 31.75 S: 2.38	X̄: 22.25 S: 6.95
Central upper lid	X̄: 15.25 S: 3.0	X̄: 16.0 S: 7.09	X̄: 16.25 S: 4.86	X̄: 32 S: 4.55	X̄: 20.5 S: 2.38

d

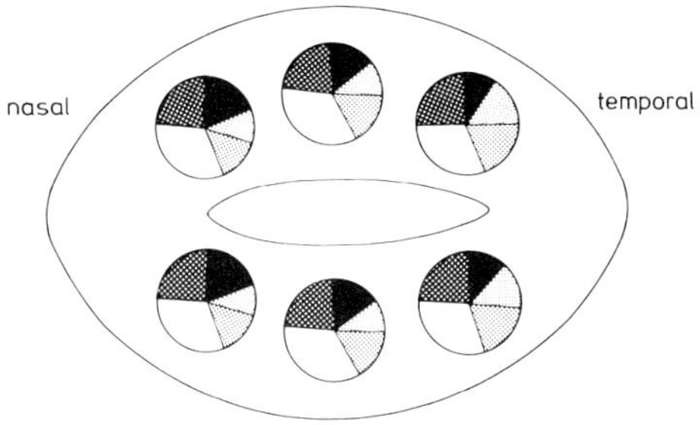

Fornix	Type I	Type II	Type III	Type IV	Type V
Nasal lower lid	$\bar{X}$: 19.31 S: 3.20	$\bar{X}$: 10.06 S: 4.64	$\bar{X}$: 14.81 S: 4.17	$\bar{X}$: 31.13 S: 3.14	$\bar{X}$: 24.69 S: 4.74
Temporal lower lid	$\bar{X}$: 11.38 S: 3.03	$\bar{X}$: 14.88 S: 3.56	$\bar{X}$: 18.69 S: 3.96	$\bar{X}$: 31.13 S: 5.28	$\bar{X}$: 23.94 S: 4.93
Central lower lid	$\bar{X}$: 14.81 S: 3.60	$\bar{X}$: 9.88 S: 4.27	$\bar{X}$: 16.44 S: 4.59	$\bar{X}$: 34.88 S: 5.42	$\bar{X}$: 24.0 S: 4.87
Nasal upper lid	$\bar{X}$: 18.63 S: 3.93	$\bar{X}$: 10.5 S: 5.11	$\bar{X}$: 14.56 S: 4.89	$\bar{X}$: 32.63 S: 6.03	$\bar{X}$: 23.69 S: 5.44
Temporal upper lid	$\bar{X}$: 9.13 S: 3.34	$\bar{X}$: 16.13 S: 5.02	$\bar{X}$: 18.88 S: 4.26	$\bar{X}$: 30.63 S: 6.36	$\bar{X}$: 24.56 S: 5.34
Central upper lid	$\bar{X}$: 13.94 S: 3.42	$\bar{X}$: 11.44 S: 6.94	$\bar{X}$: 16.88 S: 4.16	$\bar{X}$: 34.63 S: 5.63	$\bar{X}$: 23.19 S: 5.19

e

Fig. 38. Distribution of the 5 cell types in the regions studied in human conjunctival fornix in percent at various patient ages. $\bar{X}$ = Mean; S = standard deviation. *a* 1–20 years. *b* 21–40 years. *c* 41–60 years. *d* > 60 years. *e* Independently of age.

between 20 and 40 and only about 10% in subjects under 20 years of age. This third cell type was found relatively often in the lateral and central upper lid areas.

Type II epithelial cells (with secretory granules) constituted about 10% of cells in the tarsal conjunctiva. This percentage increased only in the temporal upper and lower lid areas. On the whole, they were also somewhat more frequent in older persons.

Type I (goblet) cells were more frequent (ca. 14% of epithelial surface cells) in the nasal upper and lower lids, and tended to be less frequent in the temporal lid areas. No significant changes in frequency were found in the age groups studied.

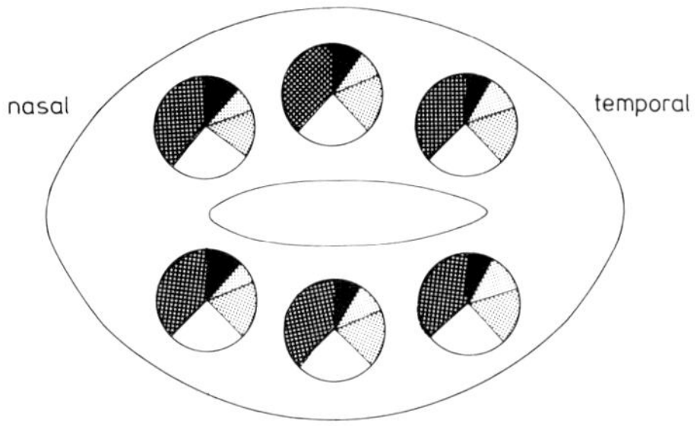

Conj. bulbi	Type I	Type II	Type III	Type IV	Type V
Upper limbus	$\bar{X}$: 0.25 S: 0.5	$\bar{X}$: 5.25 S: 4.57	$\bar{X}$: 23.75 S: 4.5	$\bar{X}$: 20.5 S: 2.08	$\bar{X}$: 50.25 S: 4.11
Nasal inferior	$\bar{X}$: 11.25 S: 0.5	$\bar{X}$: 7.5 S: 3.7	$\bar{X}$: 19 S: 3.37	$\bar{X}$: 24.5 S: 3.11	$\bar{X}$: 37.75 S: 4.11
Nasal superior	$\bar{X}$: 12.25 S: 2.75	$\bar{X}$: 6.75 S: 1.26	$\bar{X}$: 15.5 S: 0.58	$\bar{X}$: 26.25 S: 1.71	$\bar{X}$: 39.25 S: 2.63
Temporal inferior	$\bar{X}$: 7.5 S: 2.08	$\bar{X}$: 12.5 S: 2.65	$\bar{X}$: 18 S: 3.74	$\bar{X}$: 25.5 S: 4.43	$\bar{X}$: 36.5 S: 1.73
Temporal superior	$\bar{X}$: 8 S: 2	$\bar{X}$: 11.5 S: 1.29	$\bar{X}$: 18.5 S: 4.43	$\bar{X}$: 24.75 S: 2.36	$\bar{X}$: 37.25 S: 3.30
Central inferior	$\bar{X}$: 8.25 S: 1.26	$\bar{X}$: 10 S: 0.82	$\bar{X}$: 18.75 S: 5.56	$\bar{X}$: 25.5 S: 5.07	$\bar{X}$: 37.5 S: 3.32
Central superior	$\bar{X}$: 9.75 S: 2.63	$\bar{X}$: 8.75 S: 3.30	$\bar{X}$: 18.75 S: 3.30	$\bar{X}$: 24 S: 4.24	$\bar{X}$: 38.75 S: 2.22

a

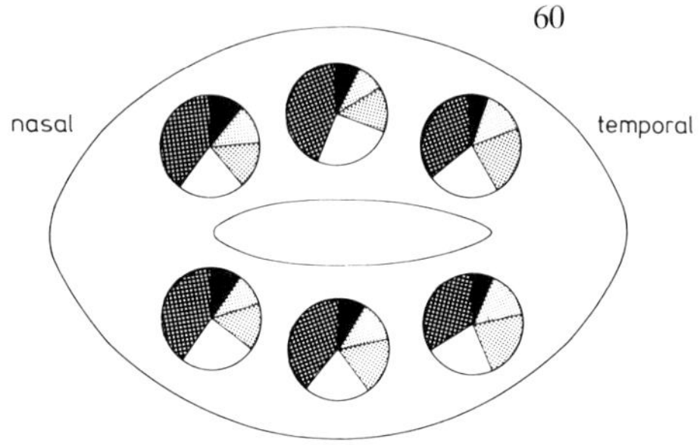

Conj. bulbi	Type I	Type II	Type III	Type IV	Type V
Upper limbus	$\bar{X}$: 0.5 S: 1	$\bar{X}$: 8.75 S: 4.65	$\bar{X}$: 20 S: 6.16	$\bar{X}$: 20 S: 3.37	$\bar{X}$: 50.75 S: 5.32
Nasal inferior	$\bar{X}$: 9.75 S: 2.22	$\bar{X}$: 10 S: 5.35	$\bar{X}$: 15.25 S: 3.86	$\bar{X}$: 24.5 S: 3.70	$\bar{X}$: 40.5 S: 5.45
Nasal superior	$\bar{X}$: 11 S: 1.41	$\bar{X}$: 12.5 S: 7.37	$\bar{X}$: 15.25 S: 8.1	$\bar{X}$: 21.25 S: 5.56	$\bar{X}$: 40 S: 6.48
Temporal inferior	$\bar{X}$: 6.5 S: 0.5	$\bar{X}$: 15.25 S: 5.62	$\bar{X}$: 21.25 S: 4.79	$\bar{X}$: 22.25 S: 3.92	$\bar{X}$: 34.75 S: 3.78
Temporal superior	$\bar{X}$: 5.5 S: 2.65	$\bar{X}$: 14 S: 2.58	$\bar{X}$: 21.75 S: 0.96	$\bar{X}$: 23 S: 1.83	$\bar{X}$: 35.75 S: 4.79
Central inferior	$\bar{X}$: 8.5 S: 3.87	$\bar{X}$: 13.25 S: 7.5	$\bar{X}$: 18 S: 0.82	$\bar{X}$: 21 S: 1.83	$\bar{X}$: 39.25 S: 4.79
Central superior	$\bar{X}$: 7 S: 2.58	$\bar{X}$: 9.75 S: 3.40	$\bar{X}$: 14.25 S: 3.78	$\bar{X}$: 24.25 S: 8.38	$\bar{X}$: 44.75 S: 4.65

b

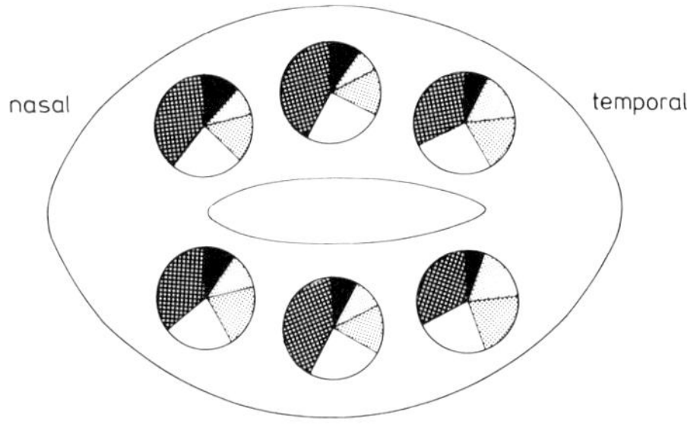

Conj. bulbi	Type I	Type II	Type III	Type IV	Type V
Upper limbus	$\bar{X}$: 0 S: 0	$\bar{X}$: 7.5 S: 6.25	$\bar{X}$: 19.5 S: 6.95	$\bar{X}$: 22 S: 7.26	$\bar{X}$: 51 S: 7.39
Nasal inferior	$\bar{X}$: 9.75 S: 3	$\bar{X}$: 11.5 S: 6.66	$\bar{X}$: 20.25 S: 3.95	$\bar{X}$: 22.5 S: 0.58	$\bar{X}$: 36 S: 3.92
Nasal superior	$\bar{X}$: 11.75 S: 2.22	$\bar{X}$: 9.25 S: 1.89	$\bar{X}$: 15.75 S: 6.55	$\bar{X}$: 23.25 S: 2.06	$\bar{X}$: 40 S: 4.0
Temporal inferior	$\bar{X}$: 5.5 S: 3.42	$\bar{X}$: 17.5 S: 7.42	$\bar{X}$: 21.25 S: 4.03	$\bar{X}$: 23 S: 2.16	$\bar{X}$: 32.75 S: 4.99
Temporal superior	$\bar{X}$: 7.5 S: 1.73	$\bar{X}$: 15.5 S: 4.93	$\bar{X}$: 18 S: 5.16	$\bar{X}$: 26.5 S: 6.25	$\bar{X}$: 32.5 S: 4.12
Central inferior	$\bar{X}$: 8 S: 1.16	$\bar{X}$: 9.25 S: 5.48	$\bar{X}$: 15.75 S: 2.63	$\bar{X}$: 24.5 S: 4.99	$\bar{X}$: 42.5 S: 3.87
Central superior	$\bar{X}$: 9.75 S: 2.63	$\bar{X}$: 7.5 S: 3.11	$\bar{X}$: 15 S: 8.45	$\bar{X}$: 25 S: 2.94	$\bar{X}$: 42.75 S: 2.22

c

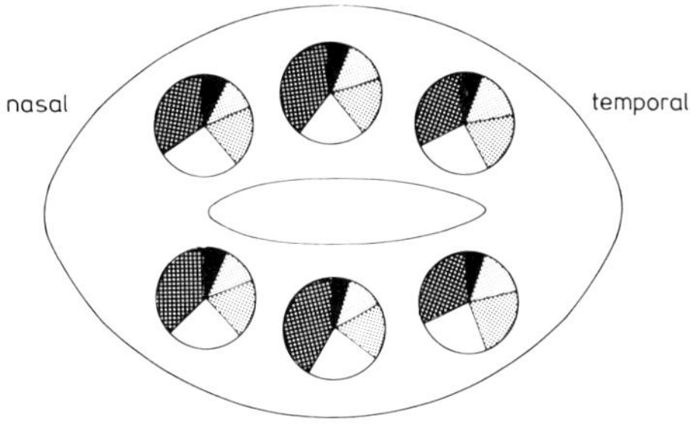

Conj. bulbi	Type I	Type II	Type III	Type IV	Type V
Upper limbus	$\bar{X}$: 0 S: 0	$\bar{X}$: 7.25 S: 3.69	$\bar{X}$: 19.5 S: 5.07	$\bar{X}$: 22 S: 2.16	$\bar{X}$: 51.25 S: 6.70
Nasal inferior	$\bar{X}$: 6.5 S: 1.0	$\bar{X}$: 12.5 S: 3.70	$\bar{X}$: 19.5 S: 2.52	$\bar{X}$: 24 S: 0.82	$\bar{X}$: 37.5 S: 4.93
Nasal superior	$\bar{X}$: 7.25 S: 3.59	$\bar{X}$: 11 S: 3.56	$\bar{X}$: 20.25 S: 1.26	$\bar{X}$: 26.75 S: 4.79	$\bar{X}$: 34.75 S: 3.86
Temporal inferior	$\bar{X}$: 5.25 S: 2.06	$\bar{X}$: 16.5 S: 4.80	$\bar{X}$: 22.5 S: 4.20	$\bar{X}$: 23.75 S: 5.56	$\bar{X}$: 32 S: 3.56
Temporal superior	$\bar{X}$: 5 S: 2.45	$\bar{X}$: 17.5 S: 2.89	$\bar{X}$: 19.75 S: 2.06	$\bar{X}$: 25.25 S: 4.35	$\bar{X}$: 32.5 S: 6.95
Central inferior	$\bar{X}$: 4.75 S: 2.06	$\bar{X}$: 12.25 S: 3.86	$\bar{X}$: 17.75 S: 6.19	$\bar{X}$: 23.25 S: 4.65	$\bar{X}$: 42 S: 5.78
Central superior	$\bar{X}$: 6 S: 2.16	$\bar{X}$: 14.25 S: 4.57	$\bar{X}$: 19 S: 5.10	$\bar{X}$: 21.25 S: 1.89	$\bar{X}$: 39.5 S: 4.80

d

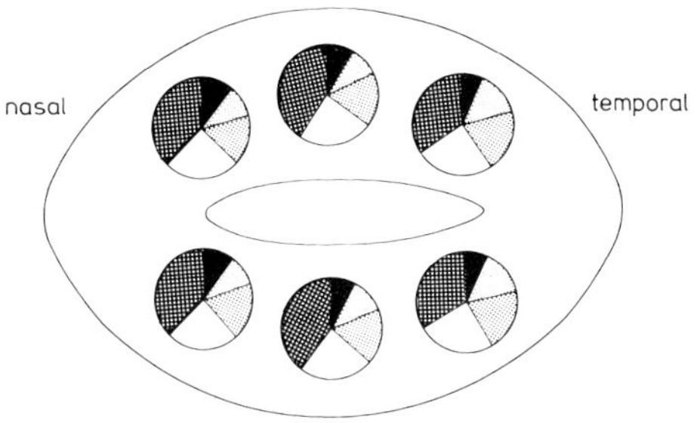

Conj. bulbi	Type I	Type II	Type III	Type IV	Type V
Upper limbus	X̄: 0.19 S: 0.54	X̄: 7.19 S: 4.55	X̄: 20.69 S: 5.46	X̄: 21.13 S: 4.06	X̄: 50.81 S: 5.39
Nasal inferior	X̄: 9.31 S: 2.30	X̄: 10.38 S: 4.88	X̄: 18.50 S: 3.69	X̄: 23.88 S: 2.36	X̄: 37.94 S: 4.48
Nasal superior	X̄: 10.56 S: 3.10	X̄: 9.88 S: 4.40	X̄: 16.69 S: 5.16	X̄: 24.38 S: 4.19	X̄: 38.5 S: 4.59
Temporal inferior	X̄: 6.12 S: 2.22	X̄: 15.44 S: 5.20	X̄: 19.56 S: 6.25	X̄: 23.56 S: 3.98	X̄: 34.0 S: 3.78
Temporal superior	X̄: 6.50 S: 2.39	X̄: 14.63 S: 3.65	X̄: 19.50 S: 3.54	X̄: 24.88 S: 3.88	X̄: 34.5 S: 4.86
Central inferior	X̄: 7.38 S: 2.63	X̄: 11.13 S: 4.84	X̄: 17.56 S: 4.08	X̄: 23.50 S: 4.24	X̄: 40.31 S: 4.57
Central superior	X̄: 8.13 S: 2.83	X̄: 10.06 S: 4.19	X̄: 16.75 S: 5.42	X̄: 23.63 S: 4.72	X̄: 41.44 S: 4.15

e

Fig. 39. Distribution of the 5 cell types in the regions studied in human bulbar conjunctiva in percent at various patient ages. X̄ = Mean; S = standard deviation. The upper limbus region is not represented schematically. *a* 1–20 years. *b* 21–40 years. *c* 41–60 years. *d* > 60 years. *e* Independently of age.

4.2.1.2. Conjunctival Fornix (fig. 38a–e)

In the fornix area, type IV cells (much rER) were the most frequent among the epithelial surface cells. This cell type was more frequent in middle age (20–60 years) than among young people or the elderly. The cells were slightly more frequent in the central conjunctival fornix, where nearly 40% of epithelial cells have abundant rER.

Epithelial cells with many mitochondria constituted somewhat less than 25% of cells in the conjunctival fornix. They were somewhat less frequent among those over 60, with just over 20%. This cell type was distributed relatively evenly over the individual regions of all age groups.

Cells with a well-developed Golgi system were, at just under 20%, the third most common cell group among epithelial surface cells of the

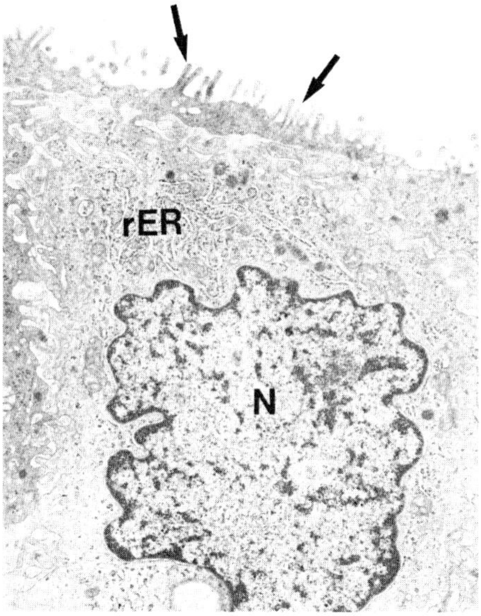

Fig. 40. Electron micrograph of a type IV cell of the tarsal conjunctiva (upper lid, temporally). N = Cell nucleus; rER = rough endoplasmatic reticulum; arrows = microvilli. Patient G.B., 48 years. × 7,050.

conjunctival fornix. They appeared less frequently in younger persons (up to 40 years old) and somewhat more often (just over 20%) in the elderly. They were found considerably more often in the temporal than in the nasal fornix area. This was true independently of the age of the subject examined (fig. 41).

The secretory (type II) epithelial cells were likewise more frequent temporally (about 15%) than nasally (about 10%) in all age groups. There was no age-specific distribution.

Type I (goblet) cells, on the other hand, were much more frequent nasally than temporally. In those under 20 years old, their 20% share of epithelial fornix cells was slightly higher than in other age groups (fig. 42).

4.2.1.3. Bulbar Conjunctiva and Limbus Epithelium (fig. 39a–e)

The fifth cell type was the most frequent among epithelial surface cells in the bulbar conjunctiva and the upper limbus region (see also fig. 28). Regardless of age, these cells constituted more than 50% in the upper limbus region, up to 40% of epithelial surface cells in other regions of bulbar conjunctiva, but slightly less in the temporal bulbar area (fig. 43).

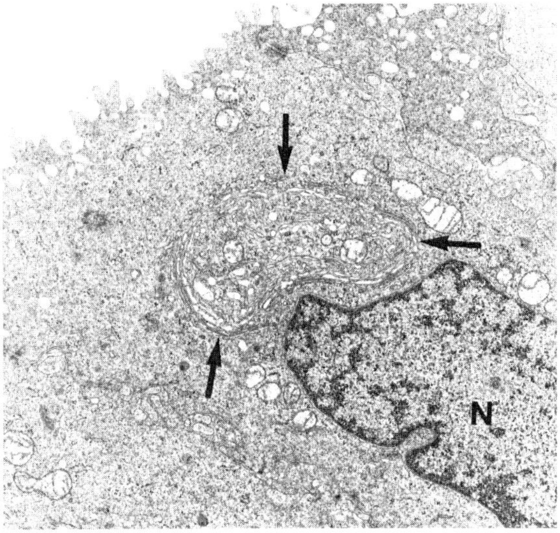

Fig. 41. Type III cell from the conjunctival fornix (central lower lid). N = Cell nucleus; arrows = Golgi complex. Patient W. Sch., 44 years. × 8,800.

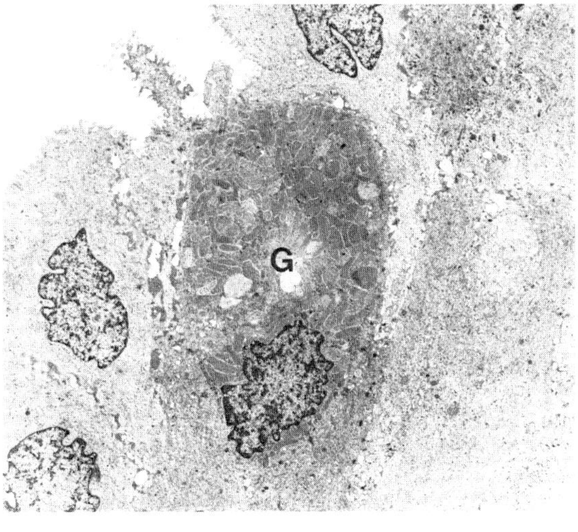

Fig. 42. Electron microscopic view of a type I goblet cell (G) of the conjunctival fornix (lateral upper lid). Patient G. D., 52 years. × 2,780.

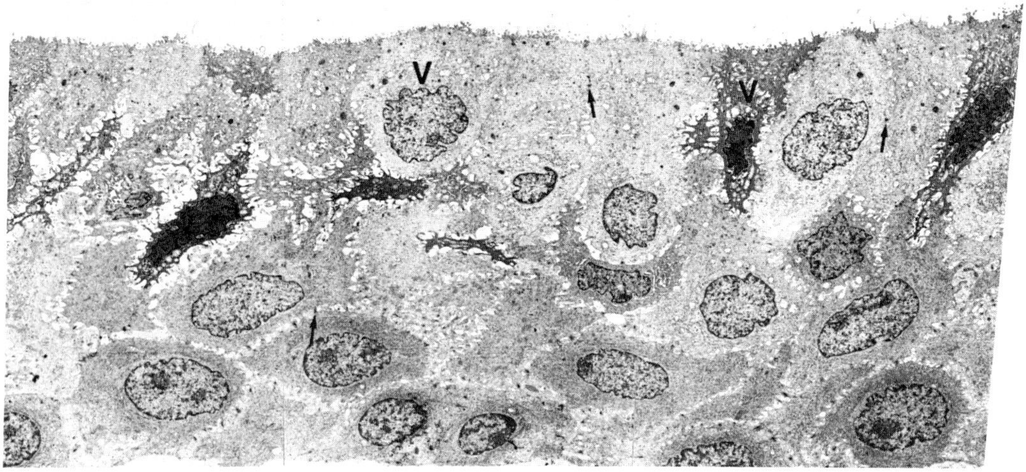

Fig. 43. Electron microscopic overview of the bulbar conjunctiva (central lower region). Note the type V cells (V), the many desmosomes (arrows) and the well-developed cytoskeleton. Patient E.G., 79 years. × 1,740.

Cells with a large amount of rER were second most frequent and constituted ca. 25% of epithelial surface cells. This cell type was more frequent in the inferior than the superior half of the bulbar conjunctiva. Its share decreased medially to a slight extent, especially in persons between 20 and 60 years old. In the limbus region, rER cells were nearly as frequent as type III (Golgi) cells, although type III cells were slightly less frequent here in those over 40. The third cell type was most frequent in temporal bulbar conjunctiva, with just under 20%. In the other regions, this cell type constituted somewhat more than 15%; the difference between temporal and nasal was more pronounced among younger subjects than older ones.

The second cell type (cells with secretory granules) was considerably more frequent in temporal bulbar conjunctiva; with 15% considerably more in the temporal superior and inferior regions than in the central (just over 10%) and nasal regions (just under 10%). This cell type was markedly less frequent among younger subjects than older ones (see fig. 39 a–d). In

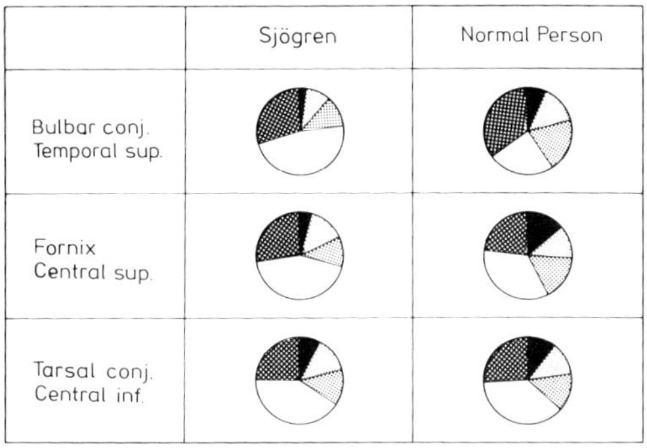

Cell types	Type I	Type II	Type III	Type IV	Type V
Bulbar conj. Temporal sup.	$\bar{X}$: 2.0 ** S: 1.90	$\bar{X}$: 9.17 S: 3.97	$\bar{X}$: 11.83 ** S: 2.71	$\bar{X}$: 46.83 ** S: 6.43	$\bar{X}$: 30.17 S: 6.56
Fornix Central sup.	$\bar{X}$: 4.5 ** S: 2.74	$\bar{X}$: 13.67 S: 4.03	$\bar{X}$: 11.00 * S: 2.97	$\bar{X}$: 43.17 * S: 4.79	$\bar{X}$: 27.67 S: 4.13
Tarsal conj. Central inf.	$\bar{X}$: 7.50 S: 3.83	$\bar{X}$: 13.67 S: 2.50	$\bar{X}$: 13.0 S: 4.05	$\bar{X}$: 41.0 S: 7.75	$\bar{X}$: 24.83 S: 5.15

Fig. 44. Distribution of the 5 cell types by percent in 3 selected regions of the conjunctival epithelium of patients with Sjögren's syndrome. $\bar{X}$ = Mean; S = standard deviation. * Significant; ** highly significant.

the limbus region about 7 of 100 epithelial surface cells belonged to the second cell type.

Goblet cells (type I) were found only very sporadically in the limbus region. In other regions of the bulbar conjunctiva their frequency lay between 5 and 10%, with nasal predilection. A slight decrease in goblet cell frequency was found here among older persons.

4.2.2. Ultrastructure of Pathological Conjunctival Tissue

4.2.2.1. Ultrastructure of Conjunctiva in Dry Eye (Sjögren's Syndrome) (fig. 44)

Independently of the evaluations of the 5 different cell types, we found a decrease in the size and number of microvilli in the tissue samples studied. Moreover, there was a pronounced desquamation of the epithelial surface cells. These desquamated epithelial cells were generally less electron-dense (fig. 45). The intercellular spaces were often expanded, with occasional infiltration of inflammatory cells (see also fig. 22). The examination of conjunctival tissue specimens for the different cell types in

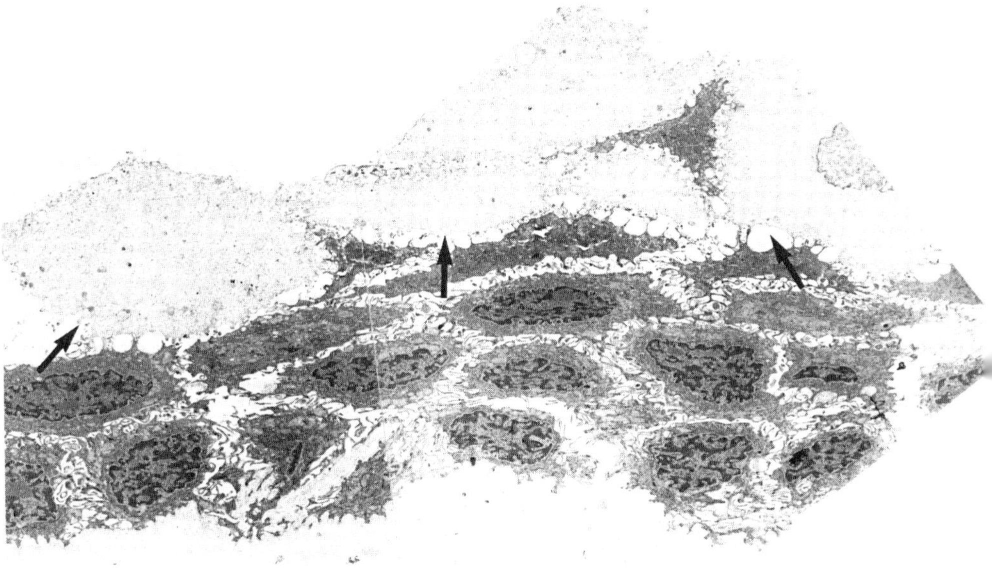

Fig. 45. Electron micrograph of the conjunctival epithelium (temporal superior bulbar conjunctiva) of a patient with Sjögren's syndrome. Note the frequent desquamation of slightly electron-dense surface epithelial cells (arrows). The intercellular reinforcements (desmosomes and tight junctions) are visibly rarefied. Patient E. L., 44 years. × 2,280.

these patients revealed that in general the cytoplasm of the surface epithelial cells has fewer organelles.

Most striking among these patients was the fact that type IV cells (with large amounts of rER) were more frequent in all regions than all other cell types (fig. 46). Their percentage was highest in the bulbar conjunctiva (20% increase), in contrast to conjunctival tissue samples from healthy subjects. The percentage of this cell type did not increase as much in the fornix or in the tarsal conjunctiva (fig. 44). The number of mitochondria-rich cells was by and large the same as that of the control group. The number of goblet cells, cells with secretory granules and type III (Golgi) cells was lower in all three regions studied, whereby the bulbar conjunctiva was more affected than the other epithelial regions.

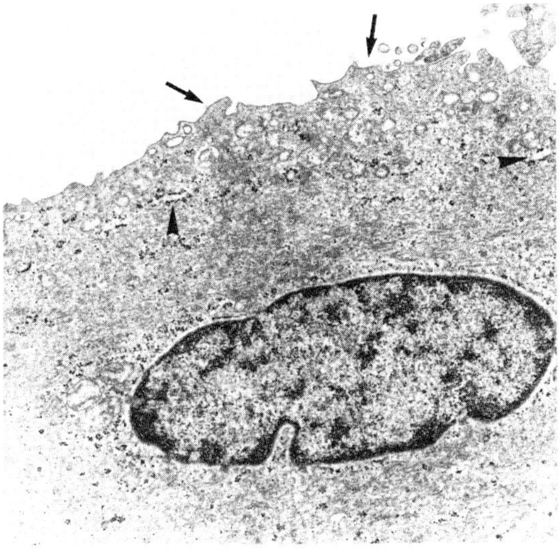

46

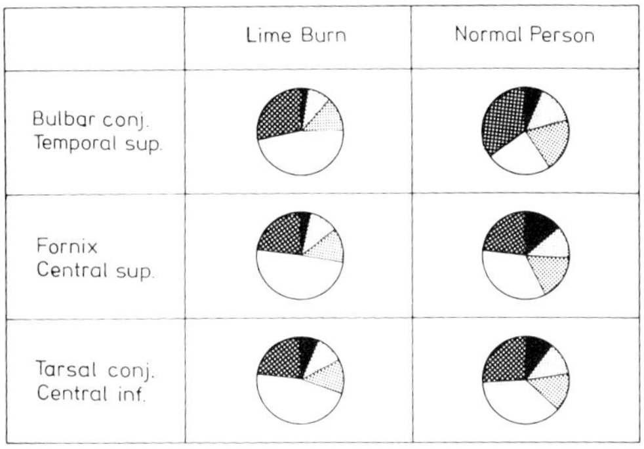

Cell types	Type I	Type II	Type III	Type IV	Type V
Bulbar conj. Temporal sup.	$\bar{X}$: 2.17 ** S: 3.49	$\bar{X}$: 9.33 S: 4.89	$\bar{X}$: 13.17 * S: 4.45	$\bar{X}$: 47.33 ** S: 8.07	$\bar{X}$: 28.33 S: 6.92
Fornix Central sup.	$\bar{X}$: 3.33 ** S: 4.84	$\bar{X}$: 11.0 S: 3.63	$\bar{X}$: 13.0 S: 4.24	$\bar{X}$: 49.67 ** S: 7.66	$\bar{X}$: 23.0 S: 4.69
Tarsal conj. Central inf.	$\bar{X}$: 6.17 S: 8.59	$\bar{X}$: 10.67 S: 4.93	$\bar{X}$: 13.50 S: 5.54	$\bar{X}$: 46.67 * S: 9.56	$\bar{X}$: 23.0 S: 2.37

47

Fig. 46. Electron micrograph of a type IV cell of tarsal conjunctiva (central lower lid) of a patient with Sjögren's syndrome. Note the reduced numbers and size of the microvilli (arrows); the cytoplasm has relatively few organelles; only scattered tubules of rER (arrowheads) are recognizable. Patient K.H., 49 years. × 7,000.

Fig. 47. Distribution of the 5 cell types by percent in 3 selected regions of the conjunctival epithelium of patients with lime burn. $\bar{X}$ = Mean; S = standard deviation. * Significant; ** highly significant.

4.2.2.2. Ultrastructure of the Conjunctiva after Lime Burn (fig. 47)

The conjunctival epithelium of this group of patients was markedly flattened in all areas. The cellular pseudoskeleton had taken over space normally occupied by the organelles (fig. 48). Distended intercellular spaces were frequent; these were filled with an amorphous exudate. In addition, some cells had intracytoplasmatic, electron-dense enclosures (fig. 49a). Occasional intraepithelial vessels were noteworthy in these chronic stages following lime burn (fig. 49b). No such changes were found among healthy subjects. Almost no goblet cells were present after lime burn, in contrast to the epithelial cells of normal conjunctiva. Only in the tarsal conjunctiva were adequate numbers of goblet cells present in 2 patients. Type II cells were still quite frequent in the conjunctival regions

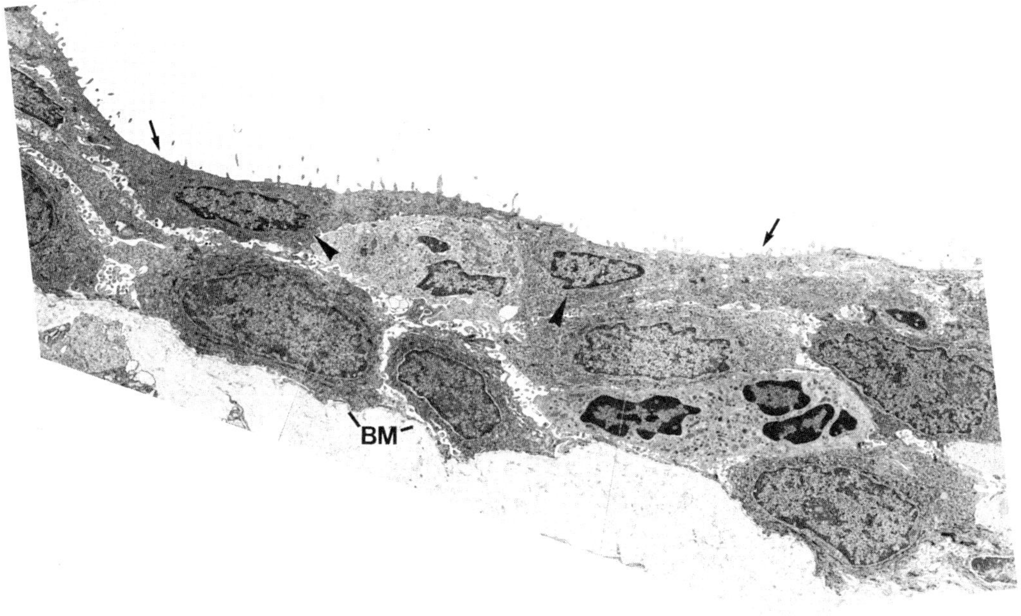

Fig. 48. Electron microscopic overview of the fornix epithelium following a lime burn. Note the much flatter surface layer of epithelial cells and the reduced size and frequency of microvilli (arrows). Goblet cells (type I) are absent; their place has been taken by numerous type IV cells (with large amounts of rER) (arrowheads). BM = Basement membrane. Patient A.B., 49 years. × 2,740.

studied; these were nearly exclusively cells with inhomogeneously osmiophilic granules. In the other surface cells, both the dominant and nondominant cell organelles were considerably less frequent than in the epithelial cells of healthy subjects. As before, this clinical picture also showed a clear-cut shift of cell types toward the type IV cells (rich in rER), which constituted almost 50% in all regions. These changes were most pronounced in the bulbar conjunctiva and the fornix. The distribution of type I and type IV cells in both epithelial regions showed highly significant differences from that of normal subjects (fig. 47).

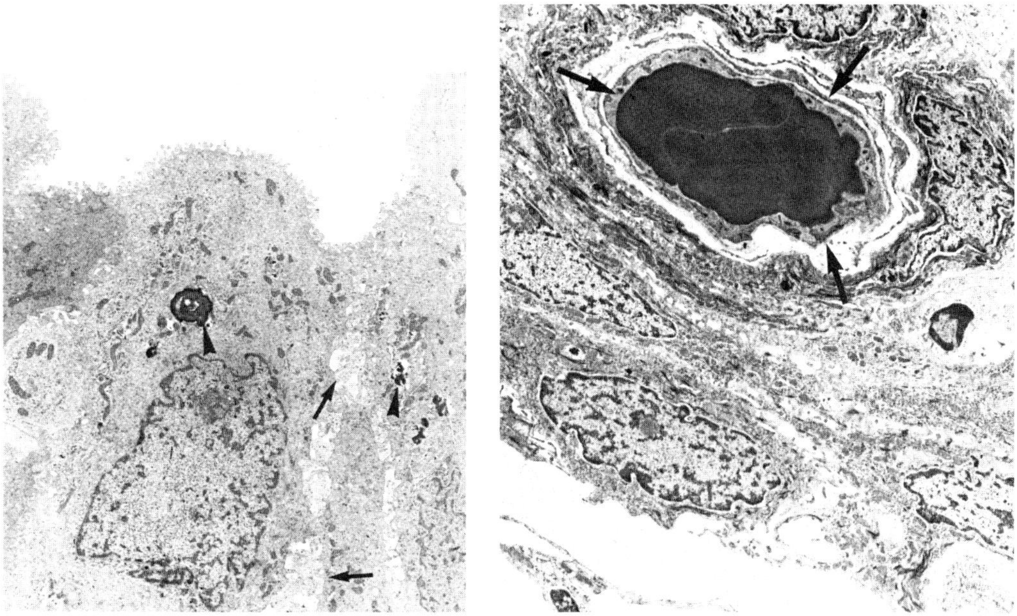

a b

Fig. 49. a Epithelial cell of tarsal conjunctiva after lime burn. Note the amorphous exudate in the intercellular spaces (arrows) and the electron-dense intracytoplasmatic enclosures (arrowheads). Patient M. G., 49 years. × 8,800. *b* Central epithelial area of the inferior tarsal conjunctiva after lime burn. Note the intraepithelial blood vessel (arrows). Patient K. R., 62 years. × 3,750.

4.2.2.3. Effect of Antiglaucomatous Agents (Pilocarpine, Timolol) on Conjunctival Ultrastructure

4.2.2.3.1. Pilocarpine (fig. 50). Desquamation of small numbers of electron-dense epithelial cells increased after the application of pilocarpine. In addition, microvilli covering the epithelium decreased in number. Somewhat less frequently several microvilli coalesced into so-called 'microplicae'. Pilocarpine had only relatively slight effect on the distribution of the 5 cell types in the regions studied. Most notable was the loss of type II cells and the increase in type IV cells. Only in bulbar conjunctiva, however, these changes were significant (type II cells) or highly significant (type IV cells) (fig. 50). Type III cells were slightly more frequent in these patients below the surface epithelial cells. As with Sjögren's syndrome and lime burns, the changes were most pronounced in bulbar conjunctiva and least pronounced in tarsal conjunctiva.

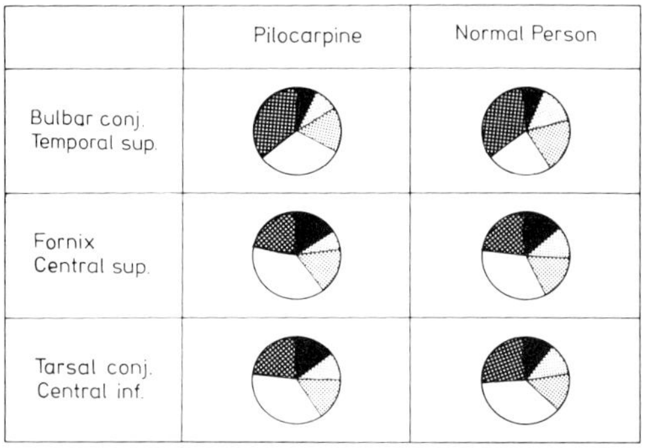

Cell types	Type I	Type II	Type III	Type IV	Type V
Bulbar conj. Temporal sup.	X̄: 6.67 S: 3.14	X̄: 9.17 * S: 2.40	X̄: 16.67 S: 3.14	X̄: 32.0** S: 2.76	X̄: 35.5 S: 4.72
Fornix Central sup.	X̄: 15.33 S: 3.08	X̄: 7.17 S: 2.23	X̄: 16.67 S: 3.72	X̄: 38.83 S: 2.79	X̄: 22.0 S: 2.0
Tarsal conj. Central inf.	X̄: 14.33 S: 2.66	X̄: 10.0 S: 3.85	X̄: 15.67 S: 2.16	X̄: 36.50 S: 3.73	X̄: 23.50 S: 2.95

50

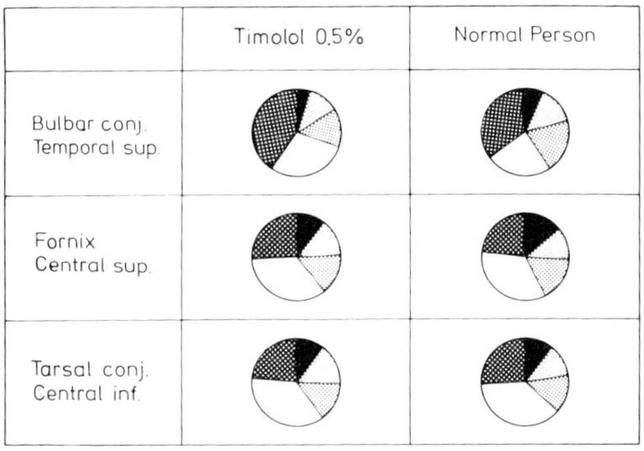

Cell types	Type I	Type II	Type III	Type IV	Type V
Bulbar conj. Temporal sup.	X̄: 4.17 S: 1.94	X̄: 11.67 S: 5.35	X̄: 14 * S: 2.10	X̄: 29.67 S: 4.23	X̄: 40.5 S: 5.43
Fornix Central sup.	X̄: 10.3 S: 1.86	X̄: 14.0 S: 5.44	X̄: 14.5 S: 3.33	X̄: 35.5 S: 3.21	X̄: 25.67 S: 2.81
Tarsal conj. Central inf.	X̄: 9.83 S: 3.87	X̄: 15.5 S: 4.32	X̄: 14.33 S: 2.66	X̄: 36.17 S: 3.19	X̄: 24.17 S: 5.57

51

Fig. 50, 51. Distribution of the 5 cell types by percent in 3 selected regions of the conjunctival epithelium of patients. X̄ = Mean; S = standard deviation. * Significant; ** highly significant. *50* After local application of pilocarpine. *51* After local timolol application.

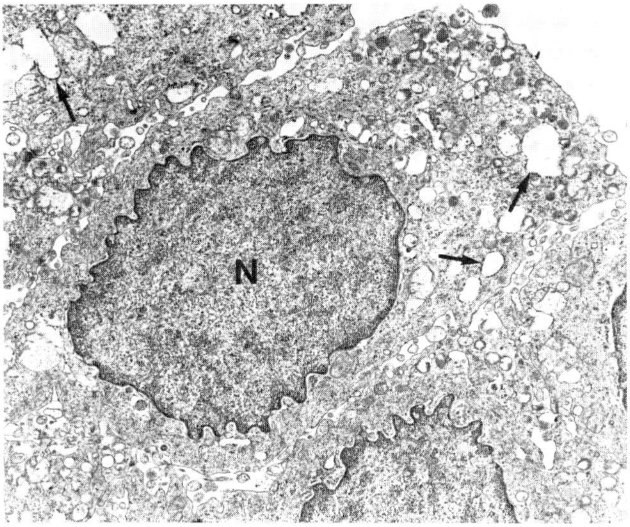

Fig. 52. Electron microscopic view of an epithelial surface cell of the conjunctival fornix of a female patient after 4 years of timolol therapy. Note the intracellular vacuolization in the apical cell area and the distended rER tubules (arrows). Tight junctions, desmosomes and microvilli are almost completely absent. N = Cell nucleus. Patient G.B., 72 years. × 7,500.

4.2.2.3.2. Timolol (fig. 51). Local application of timolol over several years was followed by degenerative changes among the microvilli of conjunctiva epithelial cells. Per cell, fewer and smaller extensions of the cytoplasm were seen on the surface. The epithelial complex was, on the whole, loosened – the desmosomes and tight junctions were considerably rarefied (fig. 52).

In addition to these extracellular changes, increased intracellular vacuolization and dilatation of rER were observed, especially in bulbar conjunctiva and the conjunctival fornix (fig. 52). The distribution of the 5 different cell types deviated most from that of normal persons in bulbar conjunctiva and the fornix, where the decrease of type I and type III cells and the increase in type IV and type V cells was greatest. In the area of the bulbar conjunctiva the change of the third cell type was significant, but the deviations of the other cell types stayed within the boundaries of a 5% probability of error. Although a tendency toward these changes was also noted in the tarsal conjunctiva, the deviations were not statistically evaluable.

5. Discussion

5.1. Morphology and Function of Rabbit and Monkey Conjunctival Epithelium

The conjunctiva's most important function – a decisive contribution to corneal transparency – is to guarantee the integrity of the tear film. In the earlier literature, final responsibility for maintaining the physiology of the tear film was assigned to only three systems, namely the Meibomian lid glands, the conjunctival goblet cells, and the lacrimal gland [38, 86, 124]. We now know, however, that a great many substances present in solution in the tear film support a complicated biological regulatory system and that the conjunctiva plays a major role in their production. The premise for this opinion, which is maintained by many authors, is derived from the ultrastructural differentiation of the conjunctival epithelium. Initial approximation reveals three different functional complexes within the conjunctival epithelium: the secretion of mucoid substances, the defence system (including regulation and self-ablution), and absorption.

5.1.1. Secretion of Mucoid Substances

Three systems participate in secreting mucoid substances. The lacrimal gland secretes chiefly glycoproteins containing sialic acid [83, 100]; the goblet cells secrete mainly mucus in daily amounts of about 2.2 μl [41]. Biochemical and histochemical analyses have shown that the secreted material of the goblet cells consists for the most part of high molecular weight sulfated and nonsulfated glycoproteins [78, 125, 161]. The glycoproteins have been further differentiated by enzymatic digestion, which showed that neuraminidase, chondro-6-sulfatase and hyaluronidase can digest most mucins [119].

Goblet cells constitute ca. 20–30% of the superficial epithelial cells of the monkey's conjunctiva. Their site of predilection is in the lower lid and the nasal and temporal upper lid. In the central upper lid, they amount to only just under 10%. In all regions studied, this first cell type is less frequent in rabbit than in monkey conjunctiva. This may be due to the fact that the more highly differentiated system of orbital glands of the lagomorpha contributes to mucin secretion, thus making some of the goblet cells dispensable [31, 206]. On the whole, goblet cells diminish in frequency from nasal to temporal.

The third system of importance for the secretion of surface-active substances is connected with the type II cells. The second cell type of human conjunctiva has been quite well described in recent times (see 5.2.). The cells draw attention in monkeys through small, electron-dense granules in the apical cytoplasm (fig. 10). They are most common in the nasal conjunctiva. Their numbers drop toward the temporal area.

This cell type appeared less often in rabbit conjunctiva, and the secretory granules were differently structured. Their ultrastructure, that is, their high osmiophily, indicates a large content of unsaturated fatty acids. Since the lipid layer of the precorneal tear film causes a sharp drop in the evaporation rate of the aqueous layer (see Introduction), it is clear that not only the orbital glands [90, 144] but also the conjunctiva contributes to the stability of this layer in animals with a very low rate of blinking.

It is well known that the secretory system of the conjunctiva reacts very sensitively to different forms of irritation [65, 94, 188]. After administering pilocarpine to monkeys for 3 months by means of a medicine dispenser, we found in quantitative electron microscopic analyses of the conjunctiva that nearly the entire goblet cell population had disappeared. Surprisingly, however, the proportion of type II cells increased from an average of just under 20% in the control animals to more than 40% in the test animals. In addition, secretory granules also increased within these cells. A well-developed rER and numerous mitochondria also indicated a very high metabolic activity (fig. 27). The morphology and distribution patterns of the other cell types, however, appeared little changed. In a control experiment in which 2 monkeys were treated with 4% pilocarpine eye drops for 3 months, we found that local application has practically no influence on the cell differentiation described earlier (fig. 25 c); this accords well with the well-known good tolerability of this medication [20, 24, 85]. The logical conclusion is that the morphological change already described was due to mechanical irritation. An increase in type II cells has been demonstrated not only in wearers of contact lenses but also in patients with trigeminus neuralgia treated with injections of alcohol into the ganglion trigeminale [34, 65].

The type III cells (chiefly characterized by a distinctive Golgi apparatus) likewise exhibited a relatively constant distribution in the animal species studied. Of all cell types in monkey conjunctiva, Golgi-rich cells were the least frequent (fig. 17); their numbers were relatively high only in the temporal upper lid (10–15%). In other areas, only ca. 5% of epithelial surface cells belonged to this cell type. In the rabbit conjunctiva, only 5% of cells at most belonged to this type, with apparent predilection for the temporal upper lid.

It is well known that polysaccharides and proteins combine to form glycoproteins within the Golgi system [43]. On the concave side of a Golgi complex, the finished product is then present in exozytotic vesicles (so-called 'condensing vacuoles') and can reach the epithelial surface, where the contents of these vesicles are released outwards through fusion of the vesicle membrane with the plasma membrane. This means that the Golgi apparatus is indispensable for synthesis of surface-active substances of the tear film. For this reason, this cell type too belongs to the functional complex which may be described as 'secretion of mucoid substances'.

5.1.2. Defense, Regulation, Self-Ablution

The particularly high proportion of type IV (rER-rich) cells in monkeys is an indicator that this cell type has special importance within the conjunctival epithelium (fig. 18). Cells which produce proteins in large quantities possess unusual amounts of rER. As described in the introduction, numerous proteins have key significance for important biological processes in the preocular tear film. The conjunctival epithelium is to be regarded as the origin of many of these substances [48, 57, 199, 210]. Histochemical methods enabled us to show that the enzyme acid phosphatase has its location within the cells of the conjunctival epithelium, indicating a high degree of metabolic activity. Moreover, a high degree of fibrinolytic activity is to be found in the area of the tarsal conjunctiva. The basis of this is a highly specialized system of plasminogen activators [160, 177].

5.1.3. Resorption

The resorption of substances always requires active transport processes, that is, energy. Only type V cells (with many mitochondria) can be adduced as a morphological basis for such processes, which may also be responsible for maintaining the ionic equilibrium of the tear film [116]. This is the second most common cell type in monkey conjunctiva and is found in slightly increased numbers in the lateral upper lid. Notably, rabbit conjunctiva exhibits this cell type most of all among the epithelial surface cells. The mitochondria-rich cells are somewhat less common in the nasal regions of the rabbit conjunctiva (fig. 19).

Using HP, we succeeded in showing a certain link between the fifth cell type and resorption processes in conjunctiva. For one thing, resorption processes were accelerated in the animal species with high numbers of mitochondria-rich epithelial cells (albino rabbits); for another, the type V cells of rabbits had already stored the HP diffusely after 5 min, those of cynomolgus monkeys on the other hand only after 30 min. Furthermore, individual cells in the tracheal epithelium of guinea pigs can store HP diffusely during the resorption process [146]. It is to be expected that these epithelial cells too will have large numbers of mitochondria. Over and above this, our experiments made it possible to describe the path of resorption for this substrate within the conjunctival epithelium.

In the initial phase of resorption, the formation of pinocytotic vesicles in the apical cytoplasm of epithelial surface cells plays a major role. These vesicles absorb some of the HP clinging to the cellular surface within 5 min (fig. 23). Of considerable importance for the enlargement of the resorbing surface are the microvilli and microplicae.

In the next phase of resorption, the vesicles penetrate the cytoplasm of the epithelial surface cells. In the process, adjacent vacuoles coalesce, leading to the formation of large transport vesicles filled with marker. These vesicles either coalesce with lysosomes, resulting in phagolysosomes in which the tracer can be digested, or release their contents (after 30 min in albino rabbits, but only after 60 min in cynomolgus monkeys) into the intercellular spaces of the epithelium (fig. 24). The finger-shaped protuberances within the intercellular spaces can function in this context either as reserve folds for the expansion of these regions or facilitate the process of exchange between intercellular and the intracellular space by enlarging the cell surface [182]. The complexes of tight junctions constitute an absolute barrier for the tracer, so that no direct connection exists between the surface of the epithelium and the intercellular spaces (fig. 20, 21). The path of HP transport therefore leads through the apical regions of the epithelial surface cells into the expandable intercellular spaces. From there a further transport of resorbed substances into the subepithelial conjunctival tissue and the abundant, fenestrated capillaries there can take place (fig. 53) [147]. Within the conjunctival epithelium, endogenic peroxidase is present both in goblet cells and in other epithelial cells [79, 80]. Nevertheless, the endogenic enzyme is easily distinguished from the marker through its location within the cell and its electron density.

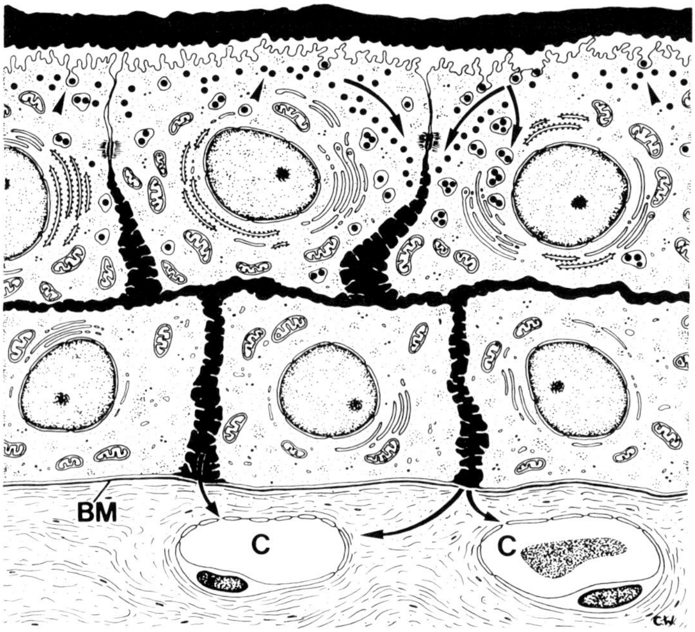

Fig. 53. Schematic view of the resorption path of HP. Reaction product first reaches the expanding intercellular spaces through the apical cytoplasmic region of the epithelial surface cells. From there, diffusion can take place into the fenestrated, subepithelial capillaries. Arrowheads = Pinocytotic vesicles; BM = basement membrane; C = subepithelial, fenestrated capillaries.

A further noteworthy finding in our study was the fact that type V cells are practically the only type present in the limbus area of albino rabbits (ca. 75%).

Inasmuch as this cell type is chiefly connected with processes requiring energy (see 5.2.), we used histochemical methods to find the enzymes ATPase and carbonic anhydrase in this epithelial area both in primates and in the albino rabbits. It is well known that these enzymes are found in epithelial cells with active functions requiring energy (e.g. in the ciliary epithelium, the epithelium of the choroidal plexus or in the renal tubules [42, 70, 107, 110]. Interestingly, both enzymes were found only among the limbal epithelial cells; neither the bordering corneal epithelium nor the epithelium of the bulbar conjunctiva were stainable. Na-K-ATPase is present in membrane-bound form in all cell layers of the limbal conjunctiva, while carbonic anhydrase is stainable both membrane-bound and cytoplasmatically only in the basal stratum [111]. We took advantage of the carbonic anhydrase content of the epithelial limbus cells and the typical morphologic characteristics of this region in order to study the regeneration of corneal epithelium of an intact limbus border. We succeeded in demonstrating that the forward-proliferating epithelial cells have already lost their initially strong stainability for carbonic anhydrase after 48 h and that only the epithelial cells proliferating forward directly from the limbus epithelium remain weakly membrane-bound and cytoplasmatically stainable (fig. 30b).

By this time the limbal epithelial cells in the middle corneal periphery had also undergone an extensive morphologic transformation into typical corneal epithelial cells (with an increase of tonofilaments, desmosomes, hemidesmosomes and tight junctions (fig. 32a, b)). Both morphologically and in localization of carbonic anhydrase, the reepithelialization was complete after 7 days.

A number of previous studies have followed the process of corneal reepithelialization from an intact limbus border. These studies paid special attention to the duration required for metaplasia of limbus epithelial cells into corneal epithelial cells according to certain criteria [98, 106, 150, 193, 207]. The criteria mentioned by the authors, however, are not specific to any one area of the epithelium of the ocular surface, but are more or less common to all epithelial cells (that is, those of the bulbar conjunctiva, the limbal epithelium and the corneal epithelium).

Only the high mitochondria content identified by us and the stainability of the limbal epithelial cells for carbonic anhydrase made it possible to study the behavior of precisely this epithelial region during reepithelialization and to determine the duration of cell transformation in light of these two criteria. Since mitochondria loss and the loss of carbonic anhydrase stainability both begin before the morphologic transformation is complete, it must be assumed that the mitochondria and the enzyme are only remnants in the migrating epithelial cells and have no special function in the forward-proliferating limbal epithelium. Since basal cells of the limbus invariably stain positive in the control preparations, we think that the intense staining of the front line of forward-proliferating epithelial cells indicates that these cells come directly from the basal layers of the limbus epithelium. The ultrastructure and histochemistry of the limbal conjunctiva both indicate that transport processes requiring energy occur in this region (see above). The purpose of these processes may be to further dehydrate the corneoscleral border area, with orientation toward the epithelial surface. Or they may have a nutritive function, meaning that they proceed in the opposite direction.

5.2. Age-Related Morphological Changes in Human Conjunctiva

The same cell types are found in human conjunctiva as those of the animal species described in section 4.1.1.1., although the distribution of individual cell types throughout the conjunctival epithelium is different, and the type II cells of human conjunctiva show more variegation. Basically, therefore, all previously described functional complexes should be human conjunctiva as well.

The type I (goblet) cells, which are responsible for mucin secretion in the lid complex, constitute ca. 15% of human epithelial surface cells. They are more common in the fornix area and especially in nasal conjunctiva (fig. 38 a–e) and show a tendency to greater frequency among younger age groups. Assuming that the average epithelial surface cell ist 10 μm wide, a 1-mm longitudinal section of epithelium will contain an average of 15 goblet cells. This varied in our findings between ± 0.54 in the limbus area and ± 5.13 in tarsal conjunctiva (central lower lid) (fig. 37e, 38e, 39e).

When excision locations are compared, these figures are slightly lower than those of Abdel-Khalek et al. [1] but considerably higher than those of Ralph [145], although the range of variation of these authors corresponded to our figures. On the other hand, our figures are considerably lower than those of Marquardt and Wenz [114] and Kessing [94] both of whom also found much more variation among their subjects than we among ours.

All these authors derived their data from light microscopic evaluation of biopsy material, a technique which includes goblet cell conglomerates not in contact with the epithelial surface in the section under study. Our criteria, however, excluded such goblet cells from consideration (see Material and Methods). Moreover, collections of goblet cells in crypts are disregarded by our evaluation technique, since crypts very rarely have complete contact with the epithelial surface in electron microscopic sections. This means that Marquardt's and Kessing's results are probably closest to the mark as far as light microscopic evaluation is concerned. In my opinion, the marked variation is due to the fact that both authors' methods included crypts in the evaluation process; these are especially numerous in the region chosen by Marquardt. As a result, one or more crypts with numerous goblet cells will be sectioned by chance in one case, while no crypts are found at the same section level in a second. Thus light microscopic evaluation of goblet cell density in conjunctiva cannot avoid the question whether crypts are present in the respective section or not.

No notable age-related differences in goblet cell density appeared in our biopsy material other than a slight increase in frequency in subjects under 20 years of age. Kessing [94] found an increased frequency of this cell type in children under 10, but this age-related peak is disputed by Marquardt [114]. Our results corresponded more to those of Marquardt, although it should be noted that our age-group classification does not take precise note of peaks in goblet cell density below 10 years of age. Abdel-Khalek et al. [1] claim that subjects over 80 have fewer goblet cells per millimeter of conjunctiva, but our experiments did not confirm this with certitude.

The second cell type (secretory cells) can be divided on the basis of our experiments into three conjunctival subgroups: cells with inhomogeneous osmiophilic vesicles are the most common and are found in all regions of conjunctiva; cells with homogeneous osmiophilic granules are much less frequent but can be present in all regions; cells with an apocrine secretion mechanism are present only among highly prismatic epithelial cells, namely in the fornix area.

The second cell type accounts for some 15% of epithelial surface cells in tarsal conjunctiva and the fornix area and about 10% in bulbar conjunctiva. It is more common temporally in all conjunctival regions. No age-related variations in frequency were ascertainable.

This second cell type of human conjunctiva, particularly the cells with inhomogeneous osmiophilic granules, have been the object of intensive research in recent years (see 1.2.3.1.). Dilly [36] found that the glycoproteins of these vesicles are not lost to the tear film; rather, after incorporation of the vesicles into the cell membrane, they cling to the membrane's outer surface and mediate the bond between the cell surface and the mucin layer of the goblet cells, thus decisively increasing the adhesiveness of the epithelial surface.

It should be cautioned, however, that in this concept the glycoproteins of the second secretory system would have no effect at the point where they are most needed, namely on the cornea. For this purpose they would have to be released to the tear film. Our findings indicate, however, that individual vesicles are released to the tear film together with their surrounding membrane, while others are incorporated into the cell membrane as Dilly describes (fig. 35b, c). Initial histochemical studies of these vesicles showed that they are stainable with colloidal iron (Hale stain method) but show no positive reaction for acid phosphatases. This means that these vesicles contain mucins and are not lyzosomes [176]. A number of authors have confirmed this conclusion [35, 67, 161].

Greiner et al. [68] showed in 1985 that the mucoproteins of this system are comprised of neutral sialo- and sulfomucins. Versura et al. [197] used lectins to show a clear-cut difference in the mucin composition of the secreta of goblet cells and type II cells. They found that sialic acid and N-acetyl-glycosamines are invariably present in the vesicles of type II cells. Moreover, N-acetyl-galactosamines are present in only a few vesicles. Mannose was never found. The mucous granules of the goblet cells, however, bind all lectins which are specific for the above-named glycoconjugates [93, 197]. It is therefore clear that the two mucous-secreting systems of conjunctiva are not interchangeable, but can maintain a functional tear film only 'in partnership'. The finding of our study that type II cells are much more frequent in temporal than in medial conjunctival regions in all age groups shows that the second cell type's secretion, important for corneal moistening, can be distributed over the cornea in adequate amounts by the medially directed wiping motion of the lids [141, 158].

The third cell type (cells with a well-developed Golgi apparatus) comprises 15% of cells in the human conjunctiva, again with strong predilection for the temporal regions of conjunctiva. In human conjunctiva, as explained in section 5.1., this cell type belongs to the functional complex 'secretion of mucoid substances'. Previous electron microscopic studies of human conjunctiva have made mention of Golgi complexes within conjunctival epithelial cells but have never systematically quantified them [1, 172, 186, 200, 202]. Since this cell type's frequency among epithelial surface cells is more or less constant, it is reasonable to assume that functions assigned specifically to the Golgi system have considerable significance for the tear film (see also 5.1.1.).

The fourth cell type (cells with large amounts of rER) is the most frequent in human conjunctiva (except for bulbar conjunctiva, where mitochondria-rich epithelial cells predominate). Some 35% of the epithelial surface cells in tarsal conjunctiva and the fornix area belong to this fourth cell type, while just under 25% of epithelial surface cells in bulbar conjunctiva are ergastoplasm-rich cells. No predilection is evident for specific regions within the conjunctiva. There also appears to be no age-related distribution; this harmonizes with the observations of Abdel-Khalek et al. [1].

Ergastoplasm-rich cells are capable of synthesizing proteins. One explanation for the proteins formed by the conjunctival epithelium is that they are carrier compounds of mucous substances, lipids, or antibodies present in the tear film. On the other hand (see Introduction), we now know that numerous enzymes and cytokines originate in the conjunctival epithelium [61, 103, 106, 111, 160, 199]. Moreover, the type IV cells, together with the mitochondria-rich cells, very likely play an important role in the process of cell proliferation, i.e. epithelial regeneration. Future studies will probably credit this cell type with even greater importance for the precorneal tear film.

Cells with numerous mitochondria are basically employed in processes requiring energy, such as the biosynthesis of lipids or proteins, cell movement, and active transport processes (secretion, resorption, active ion transport). Energy is placed at the cell's disposal by means of energy-rich adenosin triphosphate bonds. The conjunctiva requires energy in particular for processes of synthesis and transport (see above). That these functions have great importance for the conjunctival epithelium is clear from the fact that type V cells (cells with a predominance of mitochondria) are the most common in the bulbar conjunctiva and the second most common type in the fornix and the tarsal conjunctiva. In the limbus, where histochemical study of laboratory animals enabled us to identify enzymes directly involved in active transport processes (see above), this cell type constitutes by far the majority of epithelial surface cells, with just over 50%. No age-related distribution of this cell type is ascertainable. Nasal and central regions of conjunctiva exhibit somewhat more mitochondria-rich epithelial cells than the temporal regions.

The conjunctiva's resorptive capacity deserves special mention among the transport processes. This has been studied in detail with laboratory animals by us and other authors (see above). One recent study [6] even contradicts earlier literature [37] by attributing considerable importance to intraocular resorption of drugs by the conjunctiva following local application. Using radioactively labelled technetium, Sørensen and Taagehøj-Jensen [175] also demonstrated the high resorptive capacity of human conjunctiva. It may be assumed that the numerous type V cells are the morphologic basis of these active intraepithelial processes.

5.3. Effect of Pathological Processes on Cell Distribution in Human Conjunctiva

5.3.1. Conjunctival Biopsy or Impression Cytology?

All tissue specimens were taken from our subjects by conjunctival biopsy (see Material and Methods). Impression cytology, first suggested in 1977 by Egbert et al. [40], is another way of removing conjunctival material. Following refinements in the method by a Danish research group in 1980, more recent studies have attempted cytological light microscopic and SEM comparisons between pathological cell material and that of healthy subjects [60, 101, 113, 167]. Electron microscopic studies have also been carried out using material acquired by impression cytology. However, the information content of the micrographs was very small, even allowing for additional methodic difficulties [102]. The greatest disadvantage of this method is that the integrity of the epithelium is destroyed. As a result, conclusive evaluation of epithelial maturation and of surface cell and intercellular structures is not longer possible.

It must also be asked whether diagnosis with this procedure does not observe only cells released more easily from the epithelial bond due to prior damage; these cannot be representative of the epithelium under study. While it is justifiable to emphasize that impression cytology is noninvasive, the fact remains that this method must be performed under local anesthesia [60, 167]. The removal of conjunctival specimens for electron microscopic diagnosis also requires a prior local drop anesthesia (see Materials and Methods). Only pinhead-sized tissue specimens are required, however, so that bleeding and scarring can nearly always be avoided. This method of biopsy permits exact ultrastructural diagnosis with preservation of the epithelial complex. This increases information content, including that acquired in comparison of pathological and nonpathological material.

5.3.2. Dry Eye (e.g. Sjögren's Syndrome)

Keratoconjunctivitis sicca can occur as an independent pathological entity, as part of a systemic disease, as a consequence of lid anomalies (congenital, result of injury, iatrogenic) or after contact between chemical or physically noxious substances and the conjunctiva [192]. Sjögren's syndrome is an internistically defined pathological entity which can be divided into a primary form (only changes of the tear glands and salivary glands) and a secondary form (glandular involvement as a secondary consequence of systemic disease such as rheumatoid arthritis, lupus erythematodes, sclerodermia or dermatomyositis) [14]. The majority of patients with primary Sjögren's syndrome are female (9:1). At the beginning of the disease most patients are between 40 and 65 years old [14, 45]. The incidence of concomitant symptoms has been differently assessed by various authors, but keratoconjunctivitis sicca is seen in about 80% of the patients and xerostomia in about 70% [109]. Also, chronic bronchitis due to inadequate moisturization of the upper respiratory tract is found in approx. 50% of the patients [14, 45].

The presence of autoantibodies is important for diagnosis. 92% of the patients have hypergammaglobulinemia. Rheumatoid factors are found in 76% and antinuclear factors in 75%. In addition, ca. 45% of the patients have antibodies to cytoplasmic or nuclear ribonucleoproteins [9, 112]. Elevation in erythrocyte sedimentation rate, anemia and leukopenia are also frequent.

The cause of dry eye is generally regarded as infiltration of (chiefly T-) lymphocytes into the lacrimal gland, resulting in destruction of the serous glandular end sections [46]. This process certainly involves the accessory lacrimal glands as well, since our patients had a much reduced basic secretion (table 3). The electron microscope shows that the epithelial surface cells of the conjunctiva are flattened, with an increased tendency to desquamation (fig. 45). This increase in desquamation of the epithelial surface cells is in all likelihood attributable to increased mechanical pressure on the relatively dry surface of the epithelium [2, 120]. It is also possible that the morphologic changes are additionally due to the increased osmolarity of the lacrimal fluid [54]. The microvilli are reduced in quantity and size. The intercellular spaces are often distended, probably as secondary consequence of an accompanying chronic inflammatory process. Migration of inflammatory cells through the epithelium is especially common in these patients. By and large there were fewer organelles in the individual cells, in contrast to those of healthy subjects. The proportion of type V cells among epithelial surface cells of the individual regions remained unchanged. Type I (goblet) cells were much reduced in number, cells with secretory granules and type III cells less so. Cells with rER, on the other hand, were present in increased numbers (fig. 44).

Impression cytology revealed serpentine changes (so-called 'snakes'), in the chromatin of conjunctival epithelial cells of patients with keratoconjunctivitis sicca [113]. The specificity of these 'snakes' has been questioned following their recent detection in the conjunctiva of contact lens wearers

[60]. The disturbances of the tear film described by many authors are understandable in light of the large-scale ultrastructural damage to the conjunctival epithelium and should probably be viewed as a largely secondary consequence of the changed epithelial ultrastructure [32, 53, 75, 96].

5.3.3. Ultrastructure of the Conjunctival Epithelium after Lime Burn

Lime (calcium hydroxide) burn causes severe eye injury via a highly alkaline milieu (up to pH 12.4) and calcium hydroxide's exceptional ability to bind water. In the burn's initial phase, cellular elements (epithelial cells and keratocytes) are destroyed [115], and the corneal mucoproteins are altered by the deposition of calcium on the carboxyl group [191]. A few hours later, an inflammatory reaction leads to leukocytic infiltration of the burned tissue. There is also considerable exudation from distended subconjunctival capillaries [169]. Together with this cellular infiltration, the activity of lyzosomal enzymes increases [58, 137]. These glycosidases and proteases then lead to further deterioration of the corneal stroma, and corneal perforation is often unavoidable in extreme cases [148]. Over and above the size of the affected area in the corneoconjunctival epithelium [140] and the condition of the corneal basement membrane [72], dry eye caused by loss of the conjunctival goblet cells is an important factor for regeneration of the corneal epithelium in particular [145]. Among our patients, type I cells were much reduced in number in the areas studied, particularly in the bulbar conjunctiva and the fornix (fig. 47). Since our study was conducted at the chronic stage of lime burn, this loss and thus the destabilization of the tear film is to be regarded as a serious long-term complication of such burns. According to Scheffer et al. [166], the loss of goblet cells during the acute stage is due to the inflammatory tissue reaction. In the chronic stage, the regeneration of goblet cells is then inhibited by the reduced vascularization of subconjunctival scar plague.

The loss of goblet cells is generally regarded as the initial stage of conjunctival metaplasia toward epidermal differentiation. This metaplasia was recognizable in our tissue preparations by a pronounced flattening of epithelial surface cells (fig. 48). There was also a marked reduction of microvilli, as in the conjunctival specimens of patients with Sjörgen's syndrome. Type IV cells appeared ultrastructurally in greater numbers, as in Sjögren's syndrome (fig. 47). We were unable to establish whether this increased protein biosynthesis was a compensation for intensified epithelial regeneration and reduced tear secretion (the lacrimal gland's exit ducts are often occluded following lime burns) or whether more numerous type IV cells were a sensitive indicator of beginning metaplasia of nonkeratinizing squamous conjunctival epithelium. The increase of type IV cells took place chiefly at the expense of the conjunctival secretory system. Type V (mitochondria-rich) cells were less affected.

Transplantation of autologous conjunctival tissue offers some hope for partial rehabilitation of ocular epithelial cells after chemical burn [203]. It has been shown that the nonkeratinizing epithelium of the transplanted tissue is still occupied by goblet cells 28 months after transplantation [51].

5.3.4. Effect of Antiglaucomatous Drugs on the Ultrastructure of the Conjunctival Epithelium

5.3.4.1. Pilocarpine

Pilocarpine is a parasympathomimetic substance which acts through muscarine-like receptors of the postsynaptic membrane. Accordingly, only organs with parasympathetic innervation can be affected by pilocarpine. Among these in the ocular region is the lacrimal gland, whose secretion is regulated by secretomotoric nerve fibers [22, 162]. Pilocarpine is capable of raising the lacrimal gland's secretion rate 5- to 7-fold [99].

The goblet cells of the conjunctiva are not parasympathetically innervated [163]. The situation is similar in the respiratory epithelium of the trachea, where one study has conclusively shown that the subepithelially located mucus glands, not the epithelial goblet cells, are parasympathetically regulated [138]. Another study from the same research group has shown that pilocarpine activates only the secretion of the innervated subepithelial glands and that the goblet cells of the respiratory epithelium are not stimulated by pilocarpine [50].

The results of our study (see 4.1.3. and 4.2.2.3.1.) correspond to these data. We found no significant change in goblet cells after pilocarpine treatment in comparison to untreated control subjects. This finding contradicts, however, the findings of Gerstenberger and Marquardt [52]. Regarding the remaining cell types: type IV cells in particular were much increased in the bulbar conjunctiva, but there was a loss of type II cells (cells of the second conjunctival secretory system). Since these changes in conjunctival ultrastructure were quite near the lacrimal gland's efferent ducts, a direct relationship between pilocarpine and the change in rate of lacrimal gland secretion seems likely. However, the ultrastructural changes had apparently only slight influence on the tolerability of this local medication and on cell regeneration in the surface epithelium of the eye [24, 29, 85, 154].

5.3.4.2. Timolol

The local or systemic use of β-blockers has a marked effect on the tear film [17, 18]. Within the lid complex, the lacrimal gland is vulnerable to attack by adrenergic substances [5, 189]. Timolol, a β_1- and β_2-receptor blocker, lowers the lacrimal gland's secretion rate [129, 183]. Tear secretion is also reduced by timolol's anesthetic effect on the cornea, which lowers reflex secretion by the lacrimal gland [135].

In our study material there was a pronounced reduction of goblet cells and type III cells in the bulbar conjunctiva and fornix. Other cell types, particularly type IV and type V cells, increased in number. No explanation was found for these morphologic changes in the epithelial surface cells. It may be that timolol acts directly on the epithelium or the subepithelial vessels. However, the changes may also have been caused secondarily by the reduction in tear gland secretion. The changes – especially the reduction in goblet cells and the distention of the intercellular spaces – were qualitatively comparable to those of Sjögren's syndrome, even though they are much more pronounced in that disease.

Takahashi's findings, obtained in tissue cultures after the addition of timolol, show that this drug has no cytotoxic effect on conjunctival epithelial cells [187]. This indicates that the above-described morphological changes can be regarded as a secondary reaction to changes in tear secretion.

The increased desquamation of epithelial surface cells described in Results and the changed structure of the microvilli occur both in persons treated with timolol and those treated with pilocarpine. These changes may be a result of the benzalkonium chloride contained in the drops used (see Materials and Methods). It has been shown that benzalkonium chloride also affects other epithelial cells cytotoxically [24, 135, 187]. However, it is still unknown whether the increase in intracellular vacuolization and the expansion of rER in conjunctival epithelial cells of patients treated for a long time with timolol is due to timolol itself or to this preservative. On the other hand, it is possible that the reduced tear secretion of patients treated with timolol, in contrast to that of those treated with pilocarpine, heightens benzalkonium chloride's cytotoxic effect.

The changes in conjunctival epithelium just described should prompt practitioners to use timolol only with great reservation on a daily basis; moreover, patients who must be treated with this drug should undergo regular checks of tear secretion and the stability of their tear film.

6. Conclusion

Five cell types and their ultrastructural characteristics were identified in the conjunctival epithelium of different animal species (cynomolgus monkeys, albino rabbits): type I cells are goblet cells characterized in the conjunctiva by their typical form and numerous secretory enclosures; type II cells are recognizable by electron-dense granules in the apical cytoplasm; type III cells have well-developed Golgi systems; type IV cells pronounced rER, and type V cells a high mitochondria content. Each of the 5 cell types has its own characteristic pattern of distribution in the tarsal conjunctiva of cynomolgus monkeys and albino rabbits.

Whereas type I and type II cells are absent in the limbus epithelium, mitochondria-rich type V epithelial cells are very frequent there. The simultaneous presence of carbonic anhydrase and the enzyme Na-K-ATPase in the limbal epithelium makes it likely that active transport processes are at work here which depend on the transfer of ATP from the type V cells. The carbonic anhydrase content and the type V cells also proved useful for studying the transdifferentiation of limbus epithelial cells into corneal epithelial cells during reepithelialization. It was shown in the process that forward-proliferating limbal epithelial cells have already lost both of these characteristics 48 h after abrasion and accommodate themselves morphologically more and more to the corneal surface.

Our experiments with HP demonstrated a certain relationship between type V cells and resorption processes in the conjunctiva. This process was studied chronologically, and the path of resorption was traced.

The goblet cells above all (type I) and cells with secretory vesicles (type II) appear to react very sensitively to mechanical irritation. This became clearer in the tarsal conjunctiva of cynomolgus monkeys after application of pilocarpine by means of a rigid medicine dispenser.

The 5 cell types found in albino rabbits and cynomolgus monkeys were also present in the human conjunctival epithelium. However, the second cell type had 3 morphological subtypes. All 5 cell types exhibited a relatively constant pattern of distribution within the regions studied: the tarsal conjunctiva, the conjunctiva of the fornix area, and the bulbar conjunctiva (including the limbus). Surprisingly, this distribution of cell types remained largely constant even within different age groups.

Lime burns caused ultrastructural changes, most obviously in the bulbar conjunctiva and the conjunctival fornix. Most noticeable was a loss of goblet cells and an increase in type IV cells.

In dry eye (illustrated here by patients with Sjögren's syndrome), there was not only a loss of goblet cells and a decrease in other elements of the conjunctival secretory system (type II and type III cells) but also an increase in type IV cells.

Similar changes, although quantitatively much less pronounced, appeared after localized long-term treatment with the β-blocker timolol. These changes may have been due secondarily to a drop in the lacrimal gland's secretion rate, and that is why routine checking of the tear film's stability is especially to be recommended for such patients.

Changes in type II and type IV cells after long-term treatment with pilocarpine were found only in the lateral superior bulbar conjunctiva; these may also have been secondary, as a consequence of changing tear secretion (which is stimulated by pilocarpine). Benzalkonium chloride, a preservative used in eye drops, has been discussed as a possible cause of cytotoxic damage which appears both superficially after pilocarpine use and intracellularly after timolol use.

This study made it possible both to identify the morphological basis of various functional complexes of the conjunctival epithelium and, with the help of these morphological characteristics, to better classify pathological processes.

7. References

1 Abdel-Khalek, L. M. R.; Williamson, J.; Lee, W. R.: Morphological changes in the human conjunctival epithelium. I. In the normal elderly population. Br. J. Ophthal. *62:* 792–799 (1978).

2 Abdel-Khalek, L. M. R.; Williamson, J.; Lee, W. R.: Morphological changes in the human conjunctival epithelium. II. In keratoconjunctivitis sicca. Br. J. Ophthal. *62:* 800–806 (1978).

3 Abelson, M. B.; Soter, N. A.; Simon, M. A.; Dolman, J.; Allansmith, M. R.: Histamine in human tears. Am. J. Ophthal. *83:* 417–418 (1977).

4 Abelson, M. B.; Madiewale, N. A.; Weston, J. A.: The effect of topical prostaglandin D_2 on the eye. Investve Ophthal. vis. Sci. *22:* 135s (1982).

5 Aberg, G.; Adler, G.; Wikberg, J.: Inhibition and facilitation of lacrimal flow by β-adrenergic drugs. Acta ophthal. *57:* 225–235 (1979).

6 Ahmed, I.; Patton, T. F.: Importance of the noncorneal absorption route in topical ophthalmic drug delivery. Investve Ophthal. vis. Sci. *26:* 584–587 (1985).

7 Allansmith, M. R.: Defence of the ocular surface. Int. Ophtal. Clins. *12:* 93–110 (1979).

8 Allen, M.; Wright, P.; Reid, L.: The human lacrimal gland. Archs. Ophthal. *88:* 493–497 (1972).

9 Alspaugh, M. A.; Talal, N.; Tan, E. M.: Differentiation and characterization of autoantibodies and their antigens in Sjögren's syndrome. Arthritis Rheum. *19:* 216–222 (1976).

10 Andrews, J. A.: Human tear film lipids. I. Composition of the principal non-polar component. Expl. Eye Res. *10:* 223–227 (1970).

11 Arnold, R. R.; Russel, J. E.; Champion, W. J.; Brewer, M.; Gauthier, J. J.: Bactericidal activity of human lactoferrin. Infect. Immunol. *35:* 792–799 (1982).

12 Baum, J. L.: Melanocyte and Langerhans' cell population of the cornea and limbus in the albino animal. Am. J. Ophthal. *69:* 671–676 (1970).

13 Berman, M. B.; Barber, J. C.; Talamo, R. C.; Langley, C. E.: Corneal ulceration and the serum antiproteases. Investve Ophthal. vis. Sci. *12:* 759–770 (1973).

14 Bloch, K. J.; Buchanan, W. N.; Wohl, M. J.; Bunim, J. J.: Sjögren's syndrome. Medicine *44:* 187–237 (1965).

15 Bonavida, B.; Sapse, A. T.: Human tear lysozyme. II. Quantitative determination with standard Schirmer strips. Am. J. Ophthal. *66:* 70–76 (1968).

16 Bonavida, B.; Sapse, A. T.; Sercarz, E. E.: A unique lacrimal protein absent from serum and other secretions: specific tear prealbumin. Nature *221:* 375–376 (1969).

17 Bonomi, L.; Zavarise, G.; Noya, E.; Michieletto, S.: Effects of timolol maleate on tear flow in human eyes. Graefes Arch. klin. exp. Ophthal. *213:* 19–22 (1980).

18 Botelho, S.Y.; Goldstein, A.M.; Martinez, E.V.: Norepinephrine-responsive beta-adrenergic receptors in rabbit lacrimal gland. Am. J. Physiol. *224:* 1119–1122 (1973).
19 Bracciolini, M.R.: Le immunoglobuline delle lacrime. Ann. Otol. Clin. Ocul. *94:* 490–494 (1968).
20 Brewitt, H.; Dausch, D.: Untersuchungen zur Morphologie des Hornhautepithels nach Langzeitanwendung antiglaukomatöser Augentropfen. Fortschr. Ophthal. *79:* 118–124 (1982).
21 Broeckhuyse, R.M.: Tear lactoferrin: a bacteriostatic and complexing protein. Investve Ophthal. vis. Sci. *13:* 550–554 (1974).
22 Bromberg, B.B.: Autonomic control of lacrimal protein secretion. Investve Ophthal. vis. Sci. *20:* 110–116 (1981).
23 Brown, S.I.; Devichian, D.G.: The oils of the meibomian glands – physical and surface characteristics. Archs. Ophthal. *82:* 537–540 (1969).
24 Burstein, N.L.: Corneal cytotoxicity of topically applied drugs, vehicles and preservatives. Surv. Ophthal. *25:* 15–30 (1980).
25 Cardomy, P.J.; Ratazzi, M.D.; Davidson, R.G.: Tay-Sachs' disease. The use of tears for the detection of heterozygotes. New Engl. J. Med. *289:* 1072–1074 (1973).
26 Carney, L.G.; Hill, R.M.: Human tear pH. Diurnal variations. Archs. Ophthal. *94:* 821–827 (1976).
27 Carney, L.G.; Hill, R.M.: Human tear buffering capacity. Archs. Ophthal. *97:* 951–952 (1979).
28 Centifanto, Y.M.; Goorha, R.M.; Kaufman, H.E.: Interferon induction in rabbit and human tears. Am. J. Ophthal. *70:* 1006–1009 (1970).
29 Cvetković, D.; Parunović, A.; Kontić, D.J.: Konjunktivale Veränderungen bei der lokalen langjährigen medikamentösen Therapie des Glaukoms. Fortschr. Ophthal. *83:* 407–409 (1986).
30 Dark, A.J.; Durrant, T.E.; McGinty, F.; Shortland, J.R.: Tarsal conjunctiva of the upper eyelid. Am. J. Ophthal. *77:* 555–564 (1974).
31 Dart, D.; Knox, J.; Palau, A.; Botelho, S.Y.: Proteins in fluids from individual orbital glands and tears. Investve Ophthal. vis. Sci. *19:* 1342–1347 (1980).
32 Luise, V.P. de; Tabbara, K.F.: Quantitation of tear lysozyme levels in dry-eye disorders. Archs. Ophthal. *101:* 634–635 (1983).
33 Dhir, S.P.; Garg, S.K.; Sharma, Y.R.; Lath, N.K.: Prostaglandins in human tears. Am. J. Ophthal. *87:* 403–404 (1979).
34 Dilly, P.N.; Mackie, I.A.: Surface changes in the anaesthetic conjunctiva in man with special reference to the production of mucus from a non-goblet-cell source. Br. J. Ophthal. *65:* 833–842 (1981).
35 Dilly, P.N.: On the nature and role of the subsurface vesicles in the outer epithelial cells of the conjunctiva. Br. J. Ophthal. *69:* 477–481 (1985).
36 Dilly, P.N.: Contribution of the epithelium to the stability of the tear film. Trans. ophthal. Soc. UK *104:* 381–389 (1985).
37 Doan, M.G.; Jensen, A.D.; Dohlman, C.H.: Penetration routes of topically applied eye medications. Am. J. Ophthal. *85:* 383–386 (1978).
38 Dohlman, C.H.; Friend, J.; Kalevar, V.; Yagonda, D.; Balazs, E.: The glycoprotein (mucus) content of tears from normals and dry eye patients. Expl. Eye Res. *22:* 359–365 (1976).

39 Eakins, E.K.; Battacherjee, P.: Histamine, prostaglandins and ocular inflammation. Expl. Eye Res. *24:* 299–305 (1977).
40 Egbert, P.R.; Lauber, S.; Maurice, D.M.: A simple conjunctival biopsy. Am. J. Ophthal. *84:* 798–801 (1977).
41 Ehlers, N.; Kessing, S.V.; Norn, M.S.: Quantitative amounts of conjunctival mucous secretion in tears. Acta ophthal. *50:* 210–214 (1972).
42 Eichhorn, M.; Lütjen-Drecoll, E.: Vorkommen und funktionelle Bedeutung der Carboanhydrase in den Gefässendothelien. Med. Welt *36:* 697–702 (1985).
43 Farquhar, M.G.; Palade, G.E.: The Golgi apparatus – (1954–1981) – from artefact to center stage. J. Cell Biol. *91:* 77–103 s (1981).
44 Ford, C.C.; De Lange, R.J.; Petty, R.W.: Identification of a nonlysozymal bactericidal factor (beta lysin) in human tears and aqueous humor. Am. J. Ophthal. *81:* 30–33 (1976).
45 Fox, R.I.; Howell, F.V.; Bone, R.C.; Michelson, P.: Primary Sjögren syndrome: clinical and immunopathological features. Semin. Arthrit. Rheumatism *14:* 77–105 (1984).
46 Fox, R.I.; Hugli, T.E.; Lanier, L.L.; Morgan, E.L.; Howell, F.: Salivary gland lymphocytes in primary Sjögren's syndrome lack lymphocyte subsets defined by Leu-7 and Leu-11 antigens. J. Immun. *135:* 207–214 (1985).
47 Franklin, R.M.; Kenyon, K.R.; Tomasi, T.B.: Immunohistologic studies of human lacrimal gland: localisation of immunoglobulins, secretory component and lactoferrin. J. Immun. *110:* 984–992 (1973).
48 Franklin, R.M.; Prendergast, R.A.; Silverstein, A.M.: Secretory immune system of rabbit ocular adnexes. Investve Ophthal. vis. Sci. *18:* 1093–1096 (1979).
49 Gachon, A.M.; Verrelle, P.; Betail, G.; Dastugue, B.: Immunological and electrophoretic studies of human tear proteins. Expl. Eye Res. *29:* 539–553 (1979).
50 Gallagher, J.T.; Hall, R.I.; Phipps, R.J.; Jeffery, P.K.; Kent, P.W.; Richardson, P.S.: Mucus-glycoproteins (mucins) of the cat trachea: characterization and control of secretion. Biochim. biophys. Acta *886:* 243–254 (1986).
51 Geggel, H.S.; Thoft, R.A.; Friend, J.: Histology of human conjunctival transplantation. Cornea *3:* 11–15 (1984).
52 Gerstenberger, A.; Marquardt, R.: Die Becherzelldichte unter Pilocarpineinfluss. Fortschr. Ophthal. *83:* 46–50 (1986).
53 Gilbard, J.P.; Farris, R.L.; Santamaria, J.: II. Osmolarity of tears microvolumes in keratoconjunctivits sicca. Archs. Ophthal. *96:* 677–681 (1978).
54 Gilbard, J.P.; Carter, J.B.; Sang, D.N.; Refajo, M.F.; Hanninen, L.A.; Kenyon, K.R.: Morphologic effect of hyperosmolarity on rabbit corneal epithelium. Ophthalmology *91:* 1205–1212 (1984).
55 Gillette, T.E.; Allansmith, M.R.: Lactoferrin in human ocular tissues. Am. J. Ophthal. *90:* 30–37 (1980).
56 Gillette, T.E.; Greiner, J.V.; Allansmith, M.R.: Immunohistochemical localisation of human tear lysozyme. Archs Ophthal. *99:* 298–300 (1981).
57 Gluud, B.S.; Jensen, O.L.; Krogh, E.: Prostaglandingehalt (PGE_2) der Tränenflüssigkeit. Acta ophthal. *63:* 28–29s (1985).
58 Gnädinger, M.C.; Itoi, M.; Slansky, H.H.; Dohlman, C.H.: The role of collagenase in the alkali-burned cornea. Am. J. Ophthal. *68:* 478–483 (1969).
59 Goldberg, M.F.: Limbal palisades of Vogt. Trans. Am. ophthal. Soc. *8:* 155–171 (1982).
60 Götz, M.L.; Jaeger, W.; Kruse, F.E.: Die Impressionszytologie als nicht invasive

Methode der Bindehaut-Biopsie und ihre Ergebnisse. Klin. Mbl. Augenheilk. *188:* 23–28 (1986).

61 Grabner, G.; Schreiner, J.; Luger, T. A.; Stur, M.; Huber-Spitzy, V.: Hornhautepithelzellen und eine humane Bindehautzellinie (Chang) produzieren einen Interleukin 3-ähnlichen Faktor. Klin. Mbl. Augenheilk. *187:* 406–407 (1985).

62 Graumann, W.; Rohen, J. W.: Chemohistologische Befunde am menschlichen Auge (Cornea, Sklera, Uvea). Z. mikrosk.-anat. Forsch. *64:* 652–671 (1958).

63 Greiner, J. V.; Covington, H. J.; Allansmith, M. R.: Surface morphology of the human upper tarsal conjunctiva. Am. J. Ophthal. *83:* 892–905 (1977).

64 Greiner, J. V.; Gladstone, L.; Covington, H. J.; Korb, D. R.; Weidman, T. A.; Allansmith, M. R.: Branching of microvilli in the human conjunctival epithelium. Archs Ophthal. *98:* 1253–1255 (1980).

65 Greiner, J. V.; Kenyon, K. R.; Henriquez, A. S.; Korb, D. R.; Weidman, T. A.; Allansmith, M. R.: Mucus secretory vesicles in conjunctival epithelial cells of wearers of contact lenses. Archs Ophthal. *98:* 1843–1846 (1980).

66 Greiner, J. V.; Henriquez, A. S.; Covington, H. J.; Weidman, T. A.; Allansmith, M. R.: Goblet cells of the human conjunctiva. Archs Ophthal. *99:* 2190–2197 (1981).

67 Greiner, J. V.; Korb, D. R.; Allansmith, M. R.: Histochemical analysis of human non-goblet epithelial cells containing secretory vesicles. Investve Ophthal., suppl., ARVO, Sarasota 1981, p. 21.

68 Greiner, J. V.; Weidman, T. A.; Korb, D. R.; Allansmith, M. R.: Histochemical analysis of secretory vesicles in non-goblet conjunctival epithelial cells. Acta ophthal. *63:* 89–92 (1985).

69 Hansson, P. H. J.: Histochemical demonstration of carbonic anhydrase activity. Histochemie *11:* 112–118 (1967).

70 Hansson, P. H. J.: Histochemical demonstration of carbonic anhydrase activity in some epithelia noted for active transport. Acta physiol. scand. *73:* 427–434 (1968).

71 Heydenreich, A.: Die Zellimmigration im Bereich der Hornhaut. Ber. dt. ophthal. Ges. *62:* 287–294 (1959).

72 Hirst, L. W.; Fogle, J. A.; Kenyon, K. R.; Stark, W. J.: Corneal epithelial regeneration and adhesion following acid burns in the rhesus monkey. Investve Ophthal. vis. Sci. *23:* 764–773 (1982).

73 Holden, B.; Sweeney, D.; Sanderson, G.: The minimal precorneal oxygen tension to avoid corneal edema. Investve Ophthal. vis. Sci. *25:* 476–480 (1984).

74 Holly, F. J.: Formation and stability of the tear film. Int. Ophtal. Clins *13:* 73–96 (1973).

75 Holly, F. J.; Lemp, M. A.: Tear physiology and dry eyes. Surv. Ophthal. *22:* 69–87 (1977).

76 Holly, F. J.: Physical chemistry of the normal and disordered tear film. Trans. ophthal. Soc. UK *104:* 374–380 (1985).

77 Iwata, S.; Lemp, M. A.; Holly, F. J.; Dohlman, C. H.: Evaporation rate of water from the precorneal tear film and cornea in the rabbit. Investve Ophthal. vis. Sci. *8:* 613–619 (1969).

78 Iwata, S.; Kabasawa, J.: Fractionation and chemical properties of tear mucoids. Expl. Eye Res. *12:* 360–367 (1971).

79 Iwata, T.; Ohkawa, K. I.; Uyama, M.: The fine structural localisation of peroxidase activity in goblet cells of the conjunctival epithelium of rats. Investve Ophthal. vis. Sci. *15:* 40–44 (1976).

80 Iwata, T.: Cytochemical studies of endogenous peroxidase in conjunctival and corneal epithelial cells. Investve Ophthal. vis. Sci. *15:* 694–697 (1976).
81 Janssen, P.T.; Bijsterveld, O.P. van: Immunochemical determination of human tear lysozyme (muramidase) in keratoconjunctivitis sicca. Clinica chim. Acta *121:* 251–260 (1982).
82 Janssen, P.T.; Bijsterveld, O.P. van: Origin and biosynthesis of human tear fluid proteins. Investve Ophthal. vis. Sci. *24:* 623–630 (1983).
83 Jensen, O.A.; Falbe-Hansen, I.; Jacobsen, T.; Michelsen, A.: Mucosubstances of the acini of the human lacrimal gland (orbital part). I. Histochemical identification. Acta ophthal. *47:* 605–619 (1969).
84 Johnson, D.L.; Del Monte, M.A.; Cotlier, E.; Desnick, R.J.: Fabry disease: diagnosis by alpha-galactosidase activities in tears. Clinica chim. Acta *63:* 81–90 (1975).
85 Johnson, D.H.; Kenyon, K.R.; Epstein, D.L.; Buskirk, E.M. van: Corneal changes during pilocarpine gel therapy. Am. J. Ophthal. *101:* 13–15 (1986).
86 Jones, L.T.: The lacrimal secretory system and its treatment. Am. J. Ophthal. *62:* 47–60 (1966).
87 Jones, L.T.: Anatomy of the tear system. Int. Ophthal. Clins *13:* 3–22 (1973).
88 Josephson, A.S.; Weiner, R.S.: Studies of the proteins of lacrimal secretions. J. Immun. *100:* 1080–1092 (1968).
89 Kaercher, T.; Möbius, D.; Jaeger, W.: Schichtdickenbestimmung des Sekrets Meibomscher Drüsen unter In vitro-Bedingungen. Fortschr. Ophthal. *83:* 90–94 (1986).
90 Kahan, I.L.; Juhasz, K.; Mindszenti, S.: Lipids of the hadarian gland of rabbits. Acta biol. hung. *18:* 295–299 (1968).
91 Kari, O.; Salo, O.P.; Halmepuro, L.; Surilehto, K.: Tear histamine during allergic conjunctivitis challenge. Graefes Arch. klin. exp. Ophthal. *223:* 60–62 (1985).
92 Karnovsky, M.J.; Graham, R.C.: Peroxydasereaktion; in Geyer, Ultrahistochemie; 2. Aufl. (Fischer, Leipzig 1973).
93 Kawano, K.; Uehara, F.; Sameshima, M.; Ohba, N.: Applications of lectins for detection of goblet cell carbohydrates of the human conjunctiva. Expl. Eye Res. *38:* 439–447 (1984).
94 Kessing, S.V.: Mucous gland system of the conjunctiva. A quantitative normal anatomical study. Acta ophthal., *95:* suppl. (1968).
95 Kijlstra, A.; Jeurissen, S.H.M.; Koning, K.M.: Complement inhibitors in tears of different species. Expl. Eye Res. *38:* 57–62 (1984).
96 Kilp, H.; Schmid, E.; Poss, W.; Kirchner, L.: Changes of the tear film in dry eyes. Trans. ophthal. Soc. UK *104:* 450–451 (1985).
97 Kinoshita, S.; Kiorpes, T.C.; Friend, J.; Thoft, R.A.: Limbal epithelium in ocular surface wound healing. Investve Ophthal. vis. Sci. *23:* 73–80 (1982).
98 Kinoshita, S.K.; Friend, J.; Kiorpes, T.C.; Thoft, R.A.: Keratin-like proteins in corneal and conjunctival epithelium are different. Investve Ophthal. vis. Sci. *24:* 577–581 (1983).
99 Kirchner, C.: Untersuchungen über das Ausmass der Tränensekretion beim Menschen. Klin. Mbl. Augenheilk. *144:* 412–417 (1964).
100 Kreuger, J.; Sokoloff, N.; Botelho, S.Y.: Sialic acid in rabbit gland fluid. Investve Ophthal. vis. Sci. *15:* 479–481 (1976).
101 Kruse, F.E.; Schmitz, W.; Jaeger, W.; Götz, M.: Das rasterelektronenmikroskopische

Bild der Zellen des Konjunktivaepithels an Präparaten der Impressionszytologie. Klin. Mbl. Augenheilk. *188:* 29–32 (1986).
102 Kruse, F.E.; Rohen, J.W.; Götz, M.L.; Jaeger, W.; Schmitz, W.: Die Impressionszytologie der Bindehaut – Eine nicht invasive Untersuchungsmethode. Der Einsatz von REM und TEM an impressionszytologischen Präparaten. Fortschr. Ophthal. *83:* 39–45 (1986).
103 Lantz, E.; Andersson, A.: Release of fibrinolytic activators from the cornea and conjunctiva. Graefes Arch. klin. exp. Ophthal. *219:* 263–267 (1982).
104a Latkovic, S.; Nilsson, S.E.G.: The ultrastructure of the normal conjunctival epithelium of the guinea pig. I. The basal and intermediate layers of the perilimbal zone. Acta ophthal. *57:* 106–122 (1979).
104b Latkovic, S.; Nilsson, S.E.G.: II. The superficial layer of the perilimbal zone. Acta ophthal. *57:* 123–135 (1979).
104c Latkovic, S.; Nilsson, S.E.G.: III. The bulbar zone, the zone of the fornix and the supranodular zone. Acta ophthal. *57:* 305–320 (1979).
104d Latkovic, S.; Nilsson, S.E.G.: IV. The palpebral and the perimarginal zone. Acta ophthal. *57:* 321–335 (1979).
105 Latkovic, S.; Nilson, S.E.G.: Phagocytosis of latex microspheres by the epithelial cells of the guinea pig conjunctiva. Acta ophthal. *57:* 582–590 (1979b).
106 Liu, S.H.; Tagawa, Y.; Prendergast, R.A.; Franklin, R.M.; Silverstein, A.M.: Secretory component of IgA: a marker for differentiation of ocular epithelium. Investve Ophthal. vis. Sci. *20:* 100–109 (1981).
107 Lönnerholm, G.; Ridderstrale, Y.: Distribution of carbonic anhydrase in the frog nephron. Acta physiol. scand. *90:* 764–778 (1974).
108 Lönnerholm, G.: Carbonic anhydrase in rat liver and rabbit skeletal muscle, further evidence of the specificity of the histochemical cobald-phosphate method of Hansson. J. Histochem. Cytochem. *23:* 427–433 (1980).
109 Lund, O.E.: Conjunctivale Beteiligung bei Allgemeinerkrankungen. Fortschr. Ophthal. *83:* 389–395 (1986).
110 Lütjen-Drecoll, E.; Lönnerholm, G.: Carbonic anhydrase distribution in the rabbit eye by light and electron microscopy. Investve Ophthal. vis. Sci. *21:* 782–797 (1981).
111 Lütjen-Drecoll, E.; Steuhl, K.P.; Arnold, W.H.: Morphologische Besonderheiten der Conjunctiva bulbi; in Marquardt, Chronische Conjunctivitis – trockenes Auge, pp. 25–34 (Springer, Wien 1982).
112 Maddison, P.J.: Dry eye: autoimmunity and relationship to other systemic disease. Trans. ophthal. Soc. UK *104:* 458–461 (1985).
113 Marner, K.: Snake-like appearance of nuclear chromatin in conjunctival epithelial cells from patients with keratoconjunctivitis sicca. Acta ophthal. *58:* 849–853 (1980).
114 Marquardt, R.; Wenz, F.H.: Histologische Untersuchungen zur Becherzellzahl der menschlichen Bindehaut. Klin. Mbl. Augenheilk. *175:* 692–696 (1979).
115 Matsuda, H.; Smelzer, G.K.: Epithelium and stroma in alkali-burned corneas. Archs Ophthal. *89:* 396–401 (1973).
116 Maurice, D.M.: Electrical potential and ion transport across the conjunctiva. Expl. Eye Res. *15:* 527–532 (1973).
117 McClellan, B.H.; Whitney, C.R.; Newman, L.P.; Allansmith, M.R.: Immunoglobulins in tears. Am. J. Ophthal. *76:* 89–101 (1973).
118 McDermott, M.R.; Bienenstock, J.: Evidence for a common mucosal immunologic

system. I. Migration of B-immunoblasts into intestinal respiratory and genital tissues. J. Immun. *122:* 1892–1898 (1979).
119 McGhee, C. N. J.; Lee, W. R.: Quantification and enzymatic identification of conjunctival surface mucus. Trans. ophthal. Soc. UK *104:* 446–449 (1985).
120 Meyer, E.; Scharf, Y.; Schechner, R.; Zonis, S.; Scharf, Y.; Nahir, M.: Light and electron microscopical study of the conjunctiva in sicca syndrome. Ophthalmologica *190:* 45–51 (1985).
121 Mishima, S.; Maurice, D. M.: The oily layer of the tear film and evaporation from the corneal surface. Expl. Eye Res. *1:* 39–45 (1961).
122 Mishima, S.; Gasset, A.; Klyce, S. D.; Baum, J. L.: Determination of tear volume and tear flow. Investve Ophthal. vis. Sci. *5:* 264–276 (1966).
123 Montgomery, P. C.; Rohey, J. H.; Majumdar, A. J.; Lemaitre-Coelho, J. M.; Vaerman, J. P.; Ayyildiz, A.: Parameters influencing the expression of IgA antibodies in tears. Investve Ophthal. vis. Sci. *25:* 369–373 (1984).
124 Moore, J. C.; Tiffany, J. M.: Human ocular mucus origins and preliminary characterisation. Expl. Eye Res. *29:* 291–301 (1979).
125 Moore, J. C.; Tiffany, J. M.: Human ocular mucus. Chemical studies. Expl. Eye Res. *33:* 203–212 (1981).
126 Neurath, H.; Walsh, K. A.: Role of proteolytic enzymes in biological regulation. Proc. natn. Acad. Sci. USA *73:* 3825–3832 (1976).
127 Nichols, B. A.; Chiappino, M. L.; Dawson, C. R.: Demonstration of the mucous layer of the tear film by electron microscopy. Investve Ophthal. vis. Sci. *26:* 464–473 (1984).
128 Nicolaides, N.; Kaitaranta, J. K.; Rawda, T. N.; Macy, J. I.; Boswell, F. M.; Smith, R. E.: Meibomian gland studies: comparison of steer and human lipids. Investve Ophthal. vis. Sci. *20:* 522–536 (1981).
129 Nielsen, N. V.; Prause, J. U.; Eriksen, J. S.: Lysozyme alpha-1-antitrypsin and serum albumin in tear fluid of timolol-treated glaucoma patients with and without symptoms of dry eyes. Acta ophthal. *59:* 503–509 (1981).
130 Norn, M. S.: Tear secretion in normal eyes. Acta ophthal. *43:* 567–573 (1965).
131 Norn, M. S.: The conjunctival fluid, its height, volume, density of cells and flow. Acta ophthal. *44:* 212–222 (1966).
132 Norn, M. S.: Desiccation of the precorneal film. I. Corneal wetting time. Acta ophthal. *47:* 865–880 (1969).
133 Norn, M. S.: Dead, degenerated and living cells in the conjunctival fluid and mucous thread. Acta ophthal. *47:* 1102–1115 (1969).
134 Norn, M. S.: Semiquantative interference study of fatty layer of precorneal tear film Acta ophthal. *57:* 766–774 (1979).
135 O'Brien, W. J.; De Carlo, J. D.; Stern, M.; Hyndiuk, R. A.: Effects of timoptic on corneal reepithelialization. Archs Ophthal. *100:* 1331–1333 (1982).
136 Oppenheimer, D. R.; Palmer, E.; Wedell, G.: Nerve endings in the conjunctiva. J. Anat. *92:* 321–352 (1958).
137 Pahlitzsch, T.; Sinha, P.: The alkali burned cornea: electron microscopical, enzyme histochemical and biochemical observations. Graefes Arch. klin. exp. Ophthal. *223:* 278–286 (1985).
138 Peatfield, A. C.; Richardson, P. S.: Evidence for non-cholinergic non-adrenergic nervous control of mucus secretion into the cat trachea. J. Physiol. *342:* 335–342 (1983).

139 Pfister, R. R.: The normal surface of conjunctiva epithelium. A scanning electron microscopic study. Investve Ophthal. vis. Sci. *14:* 267–279 (1975).

140 Pfister, R. R.: The effects of chemical injury on the ocular surface. Ophthalmology *90:* 601–609 (1983).

141 Piffaretti, J. M.; Perrinjaquet, M.: Lidbewegungen bei normalem Lidschluß und Herstellung eines Modells. Klin. Mbl. Augenheilk. *184:* 474–475 (1984).

142 Podhorany, G.; Valci, L.; Sallai, S.; Feher, J.: Beitrag zur Auto- und Isotopie der Bindehaut. Anat. Anz. *125:* 37–41 (1969).

143 Puffer, M. J.; Neault, R. W.; Brubaker, R. F.: Basal precorneal tear turnover in the human eye. Am. J. Ophthal. *89:* 369–376 (1980).

144 Radominska-Pyrek, A.; Dabrowieki, Z.; Horrocks, L. A.: Synthesis and content of ether-linked glycerophospholipids in the hadarian gland of rabbits. Biochim. biophys. Acta *574:* 248–257 (1979).

145 Ralph, R. A.: Conjunctival goblet cell density in normal subjects and in dry eye syndromes. Investve Ophthal. vis. Sci. *14:* 299–302 (1975).

146 Ranga, V.; Kleinerman, J.: The effect of pilocarpine on vesicular uptake and transport of horseradish peroxidase by the guinea pig tracheal epithelium. Am. Rev. resp. Dis. *125:* 579–585 (1982).

147 Raviola, G.: Conjunctival and episcleral blood vessels are permeable to blood-borne horseradish peroxidase. Investve Ophthal. vis. Sci. *24:* 725–736 (1983).

148 Reim, M.; Schmidt-Martens, F. W.; Hörster, B.; Scheidhauer, E.: Morphologische und histochemische Befunde bei experimentellen Verätzungen. Ber. dt. ophthal. Ges. *77:* 749–760 (1980).

149 Reim, M.; Schmidt-Martens, F. W.: Behandlung von Verätzungen. Klin. Mbl. Augenheilk. *181:* 1–9 (1982).

150 Reim, M.; Conze, A.; Kaszuba, H. J.: Adenosine phosphate and glutathione levels in the regenerated corneal epithelium after abrasion and mild alkali burns. Graefes Arch. klin. exp. Ophthal. *218:* 42–45 (1982).

151 Rennert, O. M.; Kaiser, D.; Sollberger, H.; Joller-Jemelka, J.: Antiprotease activity in tears and nasal secretion. Hum. Genet. *23:* 73–77 (1974).

152 Richardson, K. C.; Jarrett, L.; Finke, E. H.: Embedding in epoxy resin for ultrathin sectioning in electron microscopy. Stain Technol *35:* 313–323 (1960).

153 Riesager, P. M.: A clinical study of the numerical variation with age of compact nerve endings in the human conjunctiva. J. Neurol. Neurosurg. Psychiat. *25:* 134–136 (1962).

154 Robin, J. B.; Kash, R. L.; Azen, S. P.; Schanzlin D. J.: Lack of effect of pilocarpine on corneal epithelial wound healing. Curr. Eye Res. *3:* 403–406 (1984).

155 Rodrigues, M. M.; Rowden, G.; Hackett, J.; Bakos, J.: Langerhans' cells in the normal conjunctiva and peripheral cornea of selected species. Investve Ophthal. vis. Sci. *32:* 759–765 (1981).

156 Rohen, J. W.: Morphologische Studien zur Funktion des Lidapparats. 4. Beitrag zur funktionellen Anatomie des Auges. Morph. Jb. *93:* 42–97 (1953).

157 Rohen, J. W.: Das Auge und seine Hilfsorgane; in Bargmann, Handbuch der mikroskopischen Anatomie des Menschen, Bd. 3, Teil 4 (Springer, Wien 1964).

158 Rohen, J. W.: Zur funktionellen Anatomie des Lidapparates. Ber. dt. Ophthalmol. Ges. *77:* 3–12 (1980).

159 Rohen, J. W.; Steuhl, K. P.: Specialized cell types and their regional distribution in the

conjunctival epithelium of the cynomolgus monkey. Graefes Arch. klin. exp. Ophthal. *218:* 59–63 (1982).

160 Rohen, J.W.; Steuhl, K.P.; Arnold, W.H.: Zur funktionellen Morphologie der Conjunktiva; in Marquardt, Chronische Conjunctivitis – trockenes Auge, pp. 5–24 (Springer, Wien 1982).

161 Rohen, J.W.: Zur funktionellen Morphologie der Conjunctiva. Fortschr. Ophthal. *83:* 13–24 (1986).

162 Ruskell, G.L.: The distribution of autonomic post-ganglionic nerve fibers to the lacrimal gland in monkeys. J. Anat. *109:* 229–242 (1971).

163 Ruskell, G.L.: Innervation of the conjunctiva. Trans. ophthal. Soc. UK *104:* 390–395 (1985).

164 Sapse, A.T.; Ivanyi, J.; Stone, W.; Bonavida, B.; Sercarz, E.E.: Tears as carriers of antibodies. I. Presence of antibodies to diverse antigens in rabbit tears. Archs Ophthal. *77:* 526–529 (1967).

165 Sapse, A.T.; Bonavida, B.; Stone, W.; Sercarz, E.E.: Proteins in human tears. I. Immunoelectrophoretic patterns. Archs Ophthal. *81:* 815–819 (1969).

166 Scheffer, C.; Tseng, G.; Hirst, L.W.; Maumenee, A.E.; Kenyon, K.R.; Sun, T.T.; Green, W.R.: Possible mechanisms for the loss of goblet cells in mucindeficient disorders. Ophthalmology *91:* 545–552 (1984).

167 Scheffer, C.; Tseng, G.: Staging of conjunctival squamous metaplasia by impression cytology. Ophthalmology *92:* 728–733 (1985).

168 Scherz, W.; Doane, M.G.; Dohlmann, C.H.: Tear volume in normal eyes and Keratoconjunctivitis sicca. Graefes Arch. klin. exp. Ophthal. *192:* 141–150 (1974).

169 Schmidt-Martens, F.W.; Busch, L.; Klaas, D.: Frühe Veränderungen der Conjunctiva nach Alkali-Verätzungen. Graefes Arch. klin. exp. Ophthal. *201:* 135–142 (1976).

170 Schreiner, J.; Grabner, G.; Luger, T.A.; Huber-Spitzy, V.; Haddad, R.; Stur, M.: Humane Hornhautfibroblasten und eine humane Bindehautzellinie (Chang) produzieren einen thymozytenaktivierenden Faktor (TAF) in vitro. Klin. Mbl. Augenheilk. *187:* 403–405 (1985).

171 Sen, D.K.; Sarin, G.S.: Immunoassay of human tear lysozyme. Am. J. Ophthal. *90:* 715–718 (1980).

172 Shibuya, Y.: Electron microscopy by ultra-thin specimens of normal human conjunctiva. I. Conjunctiva of the fornices. Acta ophthal. jap. *62:* 1204–1213 (1958).

173 Smolin, G.: The defence mechanism of the outer eye. Trans. ophthal. Soc. UK *104:* 363–366 (1985).

174 Sørensen, T.; Taagehøj-Jensen, F.: Tear flow in normal human eyes. Determination by means of radioisotope and gamma camera. Acta ophthal. *57:* 564–580 (1979).

175 Sørensen, T.; Taagehøj-Jensen, F.: Conjunctival transport of technetium-99^m pertechnetate. Acta ophthal. *57:* 691–699 (1979).

176 Srinivasan, B.D.; Worgul, B.V.; Iwamoto, T.; Merriam, G.R.: The conjunctival epithelium. II. Histochemical and ultrastructural studies on human and rat conjunctiva. Ophthalmic Res. *9:* 65–79 (1977).

177 Steuhl, K.P.; Arnold, H.W.: Elektronenmikroskopische und histochemische Untersuchung zur funktionellen Morphologie der Conjunctiva. Verh. anat. Ges. *75:* 965–966 (1981).

178 Steuhl, K.P.; Rohen, J.W.: Absorption of horseradish peroxidase by the conjunctival epithelium of monkeys and rabbits. Graefes Arch. klin. exp. Ophthal. *220:* 13–18 (1983).

179 Steuhl, K.P.; Rohen, J.W.: Cellular structure of the conjunctival epithelium of rabbits. Graefes Arch. klin. exp. Ophthal. *221:* 265–271 (1984).

180 Steuhl, K.P.; Döring, G.; Henni, A.; Thiel, H.J.; Botzenhart, K.: Relevance of host derived and bacterial factors in *Pseudomonas aeruginosa* corneal infections. Investve Ophthal. vis. Sci. *28:* 1559–1568 (1987).

181 Steuhl, K.P.; Thiel, H.J.: Histochemical and morphological study of the regenerating corneal epithelium after limbus-to-limbus denudation. Graefes Arch. klin. exp. Ophthal. *225:* 53–58 (1987).

182 Stossel, T.P.: Phagocytosis. New Engl. J. Med. *290:* part 1, pp. 717–723; part 2, pp. 774–780 (1974).

183 Strempel, I.: Immediatwirkung topischer β-Blocker auf die 'Breakuptime'. Ophthalmologica *192:* 11–16 (1986).

184 Stuchell, R.N.; Farris, R.L.; Mandell, I.D.: Basal and reflex human tear analysis. II. Chemical analysis: lactoferrin and lysozyme. Ophthalmology *88:* 858–861 (1981).

185 Subramanian, K.; Basu, P.K.; Thompson, G.A.: Post-operative tear cytology. Can. J. Ophthal. *5:* 65–67 (1970).

186 Suzuki, A.: Fine structure of the normal human conjunctiva. Electron microscopy in ultra-thin sections. Acta Soc. ophthal. jap. *60:* 441–459 (1956).

187 Takahashi, N.: A new method of evaluating quantitative time-dependent cytotoxicity of ophthalmic solutions in cell cultures. Beta-adrenergic blocking agents. Graefes Arch. klin. exp. Ophthal. *220:* 264–267 (1983).

188 Takakusaki, I.: Fine structure of the human palpebral conjunctiva with special reference to the pathological changes in vernal conjunctivitis. Archs hist. Jap. *30:* 247–282 (1969).

189 Tangkrisanavinont, V.: Adrenergic control of lacrimal secretion in rabbits. Life Sci. *34:* 2373–2378 (1984).

190 Terry, J.E.; Hill, R.M.: Human tear osmotic pressure. Diurnal variations and the closed eye. Archs Ophthal. *96:* 120–122 (1978).

191 Teterwak, J.: Acid mucopolysaccharides in the cornea damaged by calcium hydroxide. Klin. Oczna *39:* 543–548 (1969).

192 Thiel, H.-J.: Chronische Conjunktivitis – trockenes Auge; in Marquardt, Chronische Conjunktivitis – trockenes Auge, pp. 141–156 (Springer, Wien 1982).

193 Thoft, R.A.; Friend, J.: Biochemical transformation of regenerating ocular surface epithelium. Investve Ophthal. vis. Sci. *16:* 14–20 (1977).

194 Tiffany, J.M.: The role of Meibomian secretion in the tears. Trans. ophthal. Soc. UK *104:* 396–401 (1985).

195 Tsung, P.; Holly, F.J.: Protease activities in human tears. Curr. Eye Res. *1:* 351–355 (1981).

196 Versura, P.; Bonvicini, F.; Caramazza, R.; Laschi, R.: Scanning electron microscopy study of human cornea and conjunctiva in normal and various pathological conditions. Scan. Electron Microsc. *IV:* 1695–1708 (1985).

197 Versura, P.; Maltarello, M.C.; Bonvicini, F.; Caramazza, R.; Laschi, R.: Detection of mucus glycoconjugates in human conjunctiva by using the lectin colloidal gold technique in TEM. I. A quantitative study in normal subjects. Acta ophthal. *64:* 445–450 (1986).

198 Haeringen, N.J. van; Glasius, E.: Enzymes of energy-producing metabolism in human tear fluid. Expl. Eye Res. *18:* 407–409 (1974).

199 Haeringen, N.J. van: Clinical biochemistry of tears. Surv. Ophthal. *26:* 84–96 (1981).

200 Wanko, Th.; Lloyd, B.J.; Matthews, J.: The fine structure of human conjunctiva in the perilimbal zone. Investve Ophthal. vis. Sci. *3:* 285–301 (1964).

201 Watt, P.J.; Ward, K.A.: Mucosal immunity and the eye. Trans. ophthal. Soc. UK *104:* 367–373 (1985).

202 Weingeist, T.A.: The conjunctiva. Int. Ophthal. Clins *13:* 85–91 (1973).

203 Wesley, K.H.; Dougham, D.J.; Lindstrom, R.L.: Conjunctival autograft transplantation for unilateral ocular surface disease. Ophthalmology *90:* 1121–1126 (1983).

204 Wolf, E.: The mucocutaneous junction of the lid margin and the distribution of the tear fluid. Trans. ophthal. Soc. UK *66:* 291–305 (1946).

205 Yamamoto, G.K.; Allansmith, M.R.: Complement in tears from normal humans. Am. J. Ophthal. *88:* 758–763 (1979).

206 Yoshimura, M.; Sumeshima, M.: Mucosubstances of the acini of the rabbit hadarian gland: a histochemical and electron microscopic study. Nippon Ganka Gakkai Zasshi *84:* 101–112 (1980).

207 Zieske, J.D.; Gipson, I.K.: Protein synthesis during corneal epithelial wound healing. Investve Ophthal. vis. Sci. *27:* 1–7 (1986).

208 Zimianski, M.C.; Dawson, C.R.; Togni, B.: Epithelial cell phagocytosis of *Listeria monocytogenes* in the conjunctiva. Investve Ophthal. vis. Sci. *13:* 623–626 (1974).

209 Zirm, M.; Schmut, O.; Hofmann, H.: Quantitative Bestimmung der Antiproteinasen in der menschlichen Tränenflüssigkeit. Graefes Arch. klin. exp. Ophthal. *178:* 89–94 (1976).

210 Zirm, M.: Die Bedeutung der Proteaseinhibitoren in der Tränenflüssigkeit. Klin. Mbl. Augenheilk. *177:* 759–767 (1980).

Subject Index

N-Acetyl-galactosamines 80
Adenosin triphosphate bonds 81
Albumin, tear 2
Antibodies, aqueous layer 4, 5
Antiglaucomatous agents
  effect on conjunctival ultrastructure 27, 69–71, 85, 86, 88
  effect on cynomolgus monkey conjunctival ultrastructure 21, 42–44
Antimicrobial proteins, aqueous layer 4
Antiproteases 5
Apocrine secretion mode 53
Aqueous layer 2, 3
  antibodies 4, 5
  antimicrobial proteins 4
  biologically active components 5, 6
  cellular components 6
ATPase 76
Autoantibodies 83

Benzalkonium chloride 86, 88
Biopsy 82
β-Blockers 86
  ultrastructural effects 88
Bulbar conjunctiva 10
  antiglaucomatous agents effects 69
  cell distribution 62–65
  desmosomes 53–55
  epithelial cell types 55
  epithelium 50

Carbonic anhydrase 76, 77
  loss 77
  reaction 17
Catecholamines 6
Complement, activity in tear fluid 5, 6
Condensing vacuoles 74
Conjunctiva, *see also* Conjunctival epithelium, Conjunctival ultrastructure
  age-related morphological changes 78–81
  albino rabbits 18–20
    corneal reepithelialization 19, 20
    resorption processes 18, 19
    ultrastructure 18
  animal, cell types 28–39
  biopsy 82
  bulbar 50
    cell distribution 60–65
    desmosomes 53–55
    epithelial cell types 55
  cell types
    distribution 35–39
    pathological process effects 82–86
  diseases 13
  dry eye 26, 27
  fornix 50
    antiglaucomatous agent effects 71
    apical area 53
    cell distribution 58–59, 61, 62, 88
    epithelial surface cells 71
    type II cells 52
  glycoproteins 11
  healthy tissue 22
    data 24, 25
  impression cytology 82, 83
  light and transmission electron microscopy 9, 10
  lime burn, after 26, 27
    mucus-secreting system 11
  pathologically altered tissue 22–27
  removal locations 23
  removal procedures 22, 23
  resorptive capacity 81
  surface structure 8, 9
  tarsal 50, 52
    cell distribution 56, 59
    desmosomes 53–55
    type IV cells 67

transport processes 81
Conjunctival epithelium 6–8
cellular pseudoskeleton 67
cynomolgus monkeys 21, 39–41
defense, regulation, self-cleansing functions 74
distribution of cell types 43
function 10, 11
involvement in tear film composition 12, 13
light and transmission electron microscopy 9, 10
morphology and function of rabbit and monkey 72–78
mucoid substance secretion 72–74
osmiophilic granules 52
resorption processes 11, 74–78
rabbit and monkey 21, 39–41
secretions 11
surface structure 8, 9
Conjunctival sac 8
Conjunctival tissue transplantation 85
Conjunctival ultrastructure
albino rabbit 28, 29
antiglaucomatous agents effects 21, 27, 42–44, 69–71
cell types 87, 88
cynomolgus monkeys 20, 21, 28, 29
antiglaucomatous agent effects 21, 42–44
dry eye 65, 66
healthy
age-related and regional differences 50–55
bulbar and limbus epithelium 62–65
fornix 61, 62
tarsal 57–59
pathological 65–71
pilocarpine effect, cynomolgus monkey 42–44
study 12, 13
materials used 14–16
methods used 16, 17
Contact lenses, precorneal tear film damage 1
Cornea
basement membrane 84
carbonic anhydrase staining 49
epithelial cell morphology 77
forward-proliferating epithelium 48, 50
glycoproteins 79, 80
opacification with abrasion 20
reepithelialization at rabbit limbus 19, 20, 44–48
semitransparent area 19
Cynomolgus monkeys
conjunctival epithelium
apical section 39
cell type distribution 35–39, 87, 88
morphology and function 72–78
resorption processes 39–41
conjunctival ultrastructure 20, 21, 28, 29
effect of pilocarpine 42–44
Cytokines 81
conjunctival secretion 12

Defence system 72
Dermatomyositis 83
Desmosomes 50
bulbar and tarsal conjunctiva 53–55
conjunctiva 10
rabbit limbus 45
Dry eye 13
conjunctiva 26, 27
effects on conjunctival cell distribution 82–84
ultrastructural effects 65, 66, 88
Duplex layer 6

Electron microscopic studies 82
Epithelial cells
cytoplasm 55
cytoskeleton 53
distribution before and after pilocarpine application 43
ergastoplasm-rich 38, 45, 81
mitochondria-rich 39, 41
types 56, 57, *see also* specific types
distribution 55
Ergastoplasm-rich cells 38, 45, 81
Exozytotic vesicles 74

Fornix conjunctiva, *see* Conjunctiva, fornix

Glycoconjugates 80
Glycoproteins
formation in Golgi complex 74
mucin layer 2

secretion by lacrimal gland 72
type II cells 79, 80
Glycosidases 84
Goblet cells 2, 30, 87
age-related morphological changes 78, 79
antiglaucomatous agent effects 85
bulbar conjunctiva and limbus
epithelium 65
conjunctival 8, 9, 28
conjunctival fornix 62, 63
distribution 35–37
importance in mucin layer 12
lime burn, after 68
loss 84
mucoid substances 72, 73
mucous granules 80
reduction 86
stomata 29
tarsal conjunctiva 59
Golgi complexes 28
concave side 74
epithelium vesicles 52
fornix area 61, 62
glycoprotein formation 74
rabbit limbus 46
type I cells 28
type II cells 31, 45
type III cells 32
type IV cells 34
type V cells 34

Hale stain method 80
Hemidesmosomes 52
emergent 48
Henle's crypts 8
Histamine, tear 2, 4
Hydrophobic tear film coating 6
Hypergammaglobulinemia 83

Immunoglobulin A 4, 5
Immunoglobulins
lacrimal fluid 4, 5
serum 4
tear 2, 4
Impression cytology 82, 83
Infection mediators, tear 2
Inflammatory cell migration 83
Intercellular spaces, distended 83
Interferon 4
Intracytoplasmatic system 53
Intraepithelial processes, active 81
Intraepithelial vessels, after lime burn 67

Keratoconjunctivitis sicca 82, 83
Krause's accessory tear glands 4
Krause's subconjunctiva, joint function 8

Lacrimal fluid
antimicrobial proteins 4
immunoglobulins 4, 5
Lacrimal glands
accessory 8
dry eye 83
atrophy 1
joint function 8
mucins production 2
mucoid substances 72
secretion 1
volume 12
test of proper functioning 4
Lacrimation, reflex 4
Lactate 6
Lactoferrin 4
tear 2
Lagomorpha 73
Langerhans cells 10
Leukocytes 6
Lids
functional apparatus 1
glands 1
margins, epithelial cover 7
Light electron microscopy 9, 10
Limbus epithelium
cell distribution 62–65
enzymes 76
forward-proliferating 51
merging with bulbar conjunctiva 50
morphologic transformation 77
reepithelialization 77
ultrastructure and histochemistry 77
Lime burn
conjunctival effects 26, 27
ultrastructural effects 67–69, 84, 85, 88
Lipid layer 6
Lupus erythematodes 83
Lymphocytes 6
β-Lysin 2, 4
Lysozymes 4
tear 2

Manz glands 8
Meibomian gland 6
  marginal section 8
Melanocytes 10
Microplicae
  formation 52, 69
  T- or L-shaped 30, 38
Microvilli 50, 52
  coalesced 69
Mitochondria
  energy requirement 81
  fornix area 61
  limbal epithelium 77
  loss 77
  type V cells 57
Moll's glands 8
Mucin layer 2
  importance of globlet cells 12
  secretions 8
Mucins 2, 80
Mucoid substance secretion 72–74
Mucoproteins 80
  corneal 84
Mucus-secreting system 11

Na-K-ATPase 76, 77

Ocular limbus
  corneal reepithelialization 19, 20
  diagram 19
Osmiophilic granules 52
  after lime burn 68
  type II cells 53
    inhomogeneous 79

Peroxidase reaction 17
Phagolysosomes 75
Pilocarpine 88
  effect on conjunctival ultrastructure 27, 69, 85
  effect on cynomolgus monkey conjunctival ultrastructure 21, 42–44
  type II cell distribution after application 44, 45
Pinocytotic vesicles 40, 41
  formation 75
Plasma cells 6
Plasminogen activators, tear 2
Prealbumin, tear 2, 5
Precorneal tear film 1
  lipid layer 73
  schematic view 3
Preocular tear film 1
  cellular components 6
Prostaglandins, tear 2, 5
Proteases, tear 2, 5
Protein filaments 53
Proteins, tear 2

Rabbit
  conjunctival epithelium cell distribution 35–39, 87, 88
  microplicae 38
  morphology and function 72–78
  resorption processes 39–41
  conjunctival ultrastructure 18–20, 28, 29
  limbus, corneal reepithelialization 44–48
Reflex secretion 2, 12
Resorption process
  HP path 75, 76
  monkey conjunctival epithelium 74–78
    HP 40, 41
  morphological basis 39, 40
  rabbit conjunctival epithelium 18, 19, 74–78
    HP 41, 42
Rheumatoid arthritis 83
Rheumatoid factors 83
Rough endoplasmatic reticulum
  conjunctiva 10
  type IV cells 74
  well-developed 73

Scanning electron microscopy 17
  conjunctival appearance 9
Scleroderma 83
Semilunar fold 8
Sjögren's syndrome 13
  conjunctiva 26, 27
  conjunctival ultrastructure 65, 66, 84, 88
  effects on conjunctival cell distribution 82–84
  type IV cells 67

Tarsal conjunctiva
  antiglaucomatous agents effects 69, 71
  cell distribution 56–59

desmosomes 53–55
epithelium 50
microscopic overview 52
type IV cells 67
Tear albumin 2
Tear film
aqueous layer 2–6
secretion 8
conjunctival epithelium involvement in composition 12, 13
destabilization 84
disturbances, with dry eye 84
lipid layer 6
mucin layer 2, 8
precorneal 1
lipid layer 73
schematic view 3
preocular 1
cellular components 6
uniform distribution 3
volume 1
Tear fluid
antimicrobial proteins 4
complement activity 5, 6
proteins 2
reduced osmolarity 12
Tear glands, Krause's accessory 4
Tears, reflex 12
Tight junction complexes 75
Timolol, ultrastructural effects 27, 71, 86, 88
Tonofilaments
conjunctiva 10
connecting desmosomes in conjunctiva 53–55
corneal epithelium 48
rabbit limbus 45
Transferrin 4
Transmission electron microscopy 16, 17
conjunctiva 9, 10
Transport processes 81
Type I cells, *see* Goblet cells
Type II cells 28–31, 87
age-related morphological changes 78–80
antiglaucomatous agent effects 69, 85, 88
bulbar conjunctiva and limbus epithelium 64, 65
conjunctiva 54, 55
conjunctival fornix 52, 62
distribution 35–37
after pilocarpine application 44, 45
granules 31
lime burn, after 67, 68
mitochondria 31
mucoid substances 73
osmiophilic granular content 53
tarsal conjunctiva 59
Type III cells 32
age-related morphological change 80
antiglaucomatous agents effects 69, 71
bulbar conjunctiva and limbus epithelium 64
conjunctival fornix 63
distribution 35, 37
mucoid substances 73, 74
reduction 86
tarsal conjunctiva 57–59
Type IV cells 32–34
age-related morphological change 80, 81
antiglaucomatous agent effects 69, 71, 85, 88
distribution 37, 38
fornix area 61
increased 86
lime burn, after 68
rabbit limbus 45
rER content 74
tarsal conjunctiva 62
Type V cells 34, 87
age-related morphological change 81
antiglaucomatous agents effects 71
bulbar conjunctiva and limbus epithelium 62–65
distribution 37, 39
increased 86
mitochondria-rich 75
rabbit limbus 44
resorption function 74–76
tarsal conjunctiva 57

Urea, tear 6

Vogt, palisades 10

Wolfring glands 8

Zeiss glands 8